Cambridge Elements ≡

Elements in the Philosophy of Science
edited by
Jacob Stegenga
University of Cambridge

UNITY OF SCIENCE

Tuomas E. Tahko
University of Bristol

CAMBRIDGE
UNIVERSITY PRESS

CAMBRIDGE
UNIVERSITY PRESS

University Printing House, Cambridge CB2 8BS, United Kingdom

One Liberty Plaza, 20th Floor, New York, NY 10006, USA

477 Williamstown Road, Port Melbourne, VIC 3207, Australia

314–321, 3rd Floor, Plot 3, Splendor Forum, Jasola District Centre,
New Delhi – 110025, India

79 Anson Road, #06–04/06, Singapore 079906

Cambridge University Press is part of the University of Cambridge.

It furthers the University's mission by disseminating knowledge in the pursuit of
education, learning, and research at the highest international levels of excellence.

www.cambridge.org
Information on this title: www.cambridge.org/9781108713382
DOI: 10.1017/9781108581417

First published 2021

A catalogue record for this publication is available from the British Library.

ISBN 978-1-108-71338-2 Paperback
ISSN 2517-7273 (online)
ISSN 2517-7265 (print)

Unity of Science

Elements in the Philosophy of Science

DOI: 10.1017/9781108581417
First published online: January 2021

Tuomas E. Tahko
University of Bristol

Author for correspondence: Tuomas E. Tahko, tuomas.tahko@bristol.ac.uk

Abstract: Unity of science was once a very popular idea among both philosophers and scientists. But it has fallen out of fashion, largely because of its association with reductionism and the challenge posed by multiple realisation. Pluralism and the disunity of science are the new norm, and higher-level natural kinds and special science laws are considered to have an important role in scientific practice. What kind of reductionism does multiple realisability challenge? What does it take to reduce one phenomenon to another? How do we determine which kinds are natural? What is the ontological basis of unity? In this Element, Tuomas Tahko examines these questions from a contemporary perspective, after a historical overview. The upshot is that there is still value in the idea of a unity of science. We can combine a modest sense of unity with pluralism and give an ontological analysis of unity in terms of natural kind monism.

This title is also available as Open Access on Cambridge Core at http://dx.doi.org/10.1017/9781108581417

Keywords: philosophy of science, metaphysics of science, reduction, realisation, natural kinds

ISBNs: 9781108713382 (PB), 9781108581417 (OC)
ISSNs: 2517-7273 (online), 2517-7265 (print)

Contents

1 Introduction   1

2 A Historical Overview of Unity   2

3 Combining Unity and Pluralism   21

4 Unity of Science and Natural Kinds   40

References   65

1 Introduction

The title of this Element is likely to induce two common reactions. One is immediate scepticism, largely because the notion of the unity of science is often associated with reductionism, and nowadays (almost) no one likes reductionism. The other is one of historical interest, as for many, the idea of the unity of science brings to mind logical empiricism/positivism and perhaps the grandiose goal of completed science. But it is the *dis*unity of science that many authors are now interested in – pluralism is driving the contemporary philosophy of science. While this Element will touch on all these themes, its main target is elsewhere. The present author conceives of 'unity of science' as an *ontological* ideal – the thought that there is something that connects the various entities in reality, for instance, by way of one thing being composed of various other things. We can then ask a further question, for example, about whether the composed entities are reducible to their components or not. This way of thinking about unity of science clearly connects it with metaphysical themes about the structure of reality.

The concept of unity of science as an ontological ideal may be contrasted with unity of science as an *epistemic* ideal, focusing on the connections between the explanations and predicates of the scientific disciplines and scientific practice. This conception is at least partly motivated by the prospect of interdisciplinary research, for we do need to explain why it is useful to work across disciplinary boundaries. According to this line of thought, unity of science may have pragmatic or instrumental value, quite independently of reductionism. This reaction takes it that while the old reductionist connotations should be abandoned and pluralism is indeed thriving, there is nevertheless still something of value in the ideal of the unity of science.

This Element will start, in Section 2, by laying out a brief history of the unity of science and outlining the main reasons for the shift from unity to disunity and pluralism. Section 3 discusses the state-of-the-art regarding the unity of science, which is often driven by the epistemic/pragmatic model of unity. Two case studies will also be discussed, one from the biology-chemistry interface and one from the chemistry-physics interface. In Section 4, I will put forward my own conception of unity, following the ontological model.

We may ask: if the sciences are indeed disunified, then why is it possible to examine some higher-level phenomenon in terms of lower-level phenomena? Typically, the answer will have something to do with reduction – for example, we can explain some higher-level goings-on in terms of the behaviour of their *parts*. But what kind of reduction is this? Does it mean that there really is nothing going on at the higher level? Or does it merely mean that higher-level entities depend upon their parts? These are questions that will have an important bearing on unity of science.

A central claim of this Element is that there can be unity without *eliminative* reduction. In other words, even if we can give an ontological basis for the higher-level goings-on, say, in terms of their parts, this does not mean that we have to abandon the higher-level terminology – this is a form of *semantic anti-reductionism*. This still leaves open the question of the ontological basis of unity. My preferred answer to this question, to be developed primarily in Section 4, is that there is a singular ontological basis for unity in terms of *natural kinds*, which are ultimately what all the sciences are concerned with. The resulting view may be called *natural kind monism*: there is a single notion of 'natural kind' and anything falling under that notion can be defined in terms of the same general set of identity-criteria.[1] Natural kind monism may seem like a very controversial view, especially in the light of recent pluralistic accounts of kinds. The view concerns only the fundamental notion of 'natural kind', though: we need to postulate just one *fundamental* ontological category to account for natural kinds. This still allows us to accommodate plurality among higher-level kinds.

2 A Historical Overview of Unity

The notion of the unity of science is regularly connected to the notion of *reduction*. The initial thought is that the sciences can be unified into a theory of everything and that the theories within a single science, such as general relativity and quantum theory in physics, can also be unified. The goal would then be to ultimately reduce all higher-level phenomena to fundamental physics. According to this line of thought, unity of science just means that fundamental physics is what everything else is ultimately based on; the higher-level sciences are somehow *derivative*. The non-fundamental, higher-level sciences are typically called *special sciences*.

One way of understanding unity of science is in terms of the unity of the *entities* studied by the various sciences. The immediate challenge to this type of idea is the apparent plurality of higher-level entities such as molecules, biological organisms, and psychological states. How could such entities be accounted for solely in terms of entities studied in physics, such as fermions, bosons, and fields? This is where the notions of reduction and *bridge laws* come in, as the various levels of scientific discourse need to be somehow connected and one way to understand this connection is in terms of laws that 'bridge' the levels. A typical understanding of reduction is identity based.

---

[1] What is a criterion of identity? This question cannot be fully settled here, but in my view, the answer will involve giving an account of *sortal* terms such as 'cat' or 'mountain' (see Lowe 1989). So, in the present context, the thought is that natural kinds understood in the most general fashion will fall under a singular sortal term because they share their general identity criteria.

According to strict *reductionism*, phenomena in the higher-level special sciences are *identical* to some complex lower-level (physical) phenomena. This is an understanding of reduction as identity, which is what makes it 'strict'. Another traditional way of putting this is to say that the higher-level phenomena are nothing over and above the lower-level phenomena. In contrast, weaker forms of reductionism postulate relations that are weaker than identity to explain the connections between the sciences.

The origin of the reductionist conception of unity of science can be traced to the 1920s and 1930s, when the members of the Vienna Circle began writing about reduction. Rudolf Carnap paved the way for a new form of strong reduction, while Carl Hempel developed the deductive-nomological model of explanation. In addition, Otto Neurath pursued a somewhat more pragmatic approach to unity. Some of these different strands may be seen as culminating in Ernest Nagel's work on reduction (see Nagel 1961 for his most influential contribution).[2] Much that followed was, in fact, direct commentary on *Nagelian reduction*, which emphasized the (logical) derivability of one theory from another, with the help of bridge laws (see van Riel 2011). This line of thought was further developed and systematised in a famous article by Oppenheim and Putnam (1958). But it was soon discovered that the logical empiricist approach was overly ambitious, and the extreme reductionist picture fell out of fashion. By the 1970s, Jerry Fodor (1974) countered the ideal of the unity of science with his own: the *disunity of science*. This has become a new normal: almost no one now believes that we can unify the sciences in the manner suggested by the strong reductionists. But to see why this is the case, it is worthwhile to briefly examine the idea of Nagelian reduction and the work by Oppenheim and Putnam that followed. This will be covered in Sections 2.1 and 2.2, before moving on to Fodor's reaction and the debate with Jaegwon Kim that followed, in Sections 2.3 and 2.4. The phenomenon of *multiple realisability* has a key role in this debate, and the line of thought is finalised in Section 2.5 with a discussion of Louise Antony's analysis.

Before we get started, a simple figure showing the varieties of unity (Figure 1) might be helpful. This is by no means the only way to distinguish different approaches to unity, and it should be noted that there are further distinctions to be made in the various subcategories. Simplified as it is, Figure 1 may give us a useful starting point. I will not provide detailed definitions of these varieties yet, as some historical context will be needed to

---

[2] See Cat (2017) for a more comprehensive discussion of the history of the unity of science, which does, of course, go back much further than the Vienna Circle. On Nagel's view, see, for example, Needham (2010) and van Riel (2011), and on Neurath's influence, see the articles in Symons, Pombo, and Torres (2011).

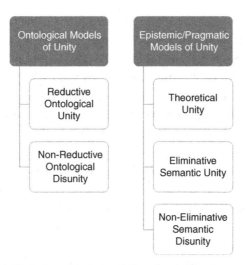

Figure 1 Varieties of unity and disunity (author's own work)

make sense of them. Accordingly, the following definitions should also be taken as tentative rather than final. Note also that the various versions of unity on the left-hand side and right-hand side are not necessarily mutually exclusive and indeed sometimes they are explicitly combined in various ways. Finally, while I have labelled two of the options as 'disunity', this does not mean that there could be no sense of unity involved – disunity merely entails a type of pluralism. The reader is invited to refer back to Figure 1 as needed.

Ontological models of unity, as the name suggests, concern the ontological structure of reality. They are intended to be objective models about how reality is structured, whether levelled or not.

Reductive ontological unity suggests that all entities reduce to some base class of entities, typically, those of fundamental physics.

Non-reductive ontological disunity suggests that reality may be structured into non-reductive levels that are connected, for instance, by compositional relations, where the composed entities are a genuine addition to reality.

Epistemic/pragmatic models of unity concern the structure of scientific theories and are hence guided by epistemic, explanatory, or pragmatic considerations relating to scientific practice.

Theoretical unity (or *unity of formalism*) suggests that a certain set of distinct phenomena may be (approximately) described in terms of a unified formal (mathematical) framework.

Eliminative semantic unity suggests that all predicates of higher-level sciences are identical to predicates of (fundamental) physics; hence all higher-level explanations are, in principle, replaceable by lower-level (physical) explanations.

Non-eliminative semantic disunity suggests that higher-level or special science predicates cannot be identified with predicates of physics (traditionally because of arguments from multiple realisation); hence higher-level explanations cannot be dispensed with and a type of pluralism is allowed. However, this view is typically combined with the idea that all higher-level predicates refer to entities that can be understood as being linked to lower-level entities (e.g., by using compositional or mechanistic explanations). This type of combination of *non-eliminative semantic disunity* and *non-reductive ontological disunity* is often called *non-reductive physicalism*.

2.1 From Logical Empiricism to Nagelian Reduction

Unity of science was a driving ideal in the logical empiricist tradition. It was closely tied to reductionism and an anti-metaphysical attitude, so the relevant sense of unity was primarily following the epistemic/pragmatic model of unity with a semantic focus on the language of science, but it may be considered to have had an implicit ontological element as well. The resulting picture is a combination of eliminative semantic unity and reductive ontological unity. We can see all these elements in Carnap's (e.g., 1928, 1934) work, and indeed in his formulation of unity in the book entitled *The Unity of Science*: 'all empirical statements can be expressed in a single language, all states of affairs are of one kind and are known by the same method' (Carnap 1934: 32).

While this formulation focuses on semantic and epistemic unity, it certainly suggests an ontological commitment to states of affairs of just one kind as well. However, I will omit an analysis of Carnap's metaphilosophical position (for a related discussion, see Tahko 2015a). The important feature here is the apparently anti-metaphysical background, which meant that unity of science was conceived as an epistemological and methodological project attempting to establish that all higher-level, special science statements, predicates, and explanations are reducible into those of physics. The implicit ontological import would then be that all *entities* also reduce to those of physics. These ideas were clearly present in Carnap's work. But instead of focusing on the well-reached history of the logical empiricist position and details of the work of its main architects, such as Carnap, we can directly jump to two of the core elements regarding the development of the logical empiricist tradition towards a systematic model of reduction and unity.

These two core elements are Hempel's (1942; Hempel and Oppenheim 1948; Hempel 1965) deductive-nomological (DN) model of explanation and Nagel's (1961) account of reduction. Both of these elements focus on explanatory connections. In brief, the DN model involves an *explanandum*, a sentence that describes the phenomenon we wish to explain, and an *explanans*, which is a class of sentences providing the explanation for the phenomenon. The deductive part of the model concerns the requirement that the explanandum must follow from the explanans by logical consequence. The nomological (i.e., 'law-like') part refers to the requirement that the explanans must contain at least one law of nature, and without it the deductive inference would not be valid. For example, take the universal generalisation that 'all gases expand when heated under constant pressure' (Hempel 1965: 338), which may be regarded as a law. According to the DN model, we can use this law and the fact that a given sample of gas has been heated under constant pressure to explain why the sample of gas has expanded.

Now let us connect the DN model with Nagelian reduction. Here is a representative passage from Nagel himself: 'A reduction is effected when the experimental laws of the secondary science...are shown to be the logical consequences of the theoretical assumptions (inclusive of the coordinating definitions) of the primary science' (Nagel 1961: 352). The basic form of a Nagelian reduction suggests that theory A reduces to theory B if and only if A is derivable from B with the help of any necessary 'coordinating definitions', which are more commonly known as *bridge laws* or bridge principles; these bridge laws can take the form of logical connections, conventions, or empirical hypotheses (1961: 354). It is not difficult to see that the DN model and Nagelian reduction are closely connected. In effect, the Nagelian model suggests that the reducing theory explains the reduced theory (with the help of bridge laws).

The DN model and Nagelian reduction dominated the philosophy of science for a time, but they did also come under heavy criticism (see, e.g., van Riel 2014). We will not need to provide these details here, but it is important that we understand the sense of unity that emerges from this background: the upshot is that all higher-level science explanations ultimately collapse into those of physics. But while the DN model and Nagelian reduction both operate at the epistemic level of explanation, there does seem to be an implicit ontological commitment to the idea that all entities reduce to (fundamental) physical entities. However, the difference between these two aspects needs to be clarified. In Oppenheim and Putnam (1958), which is the subject of the next section, we can find a clearer statement of the ontological view.

2.2 Oppenheim and Putnam on the Unity of Science

Oppenheim and Putnam (1958) distinguish three different unity theses:

- Unity of language (epistemological reduction)
- Unity of laws (implies the unity of language)
- Unity of science (in the strongest sense, implies unity of laws and unity within a science)

Unity of language is described as the idea that all *terms of science* are reduced to one discipline, such as physics, whereby 'reduction' would most plausibly be understood definitionally. This would result in a type of *epistemological reduction*, where all the claims of the special sciences could, at least in principle, be translated into claims of some more fundamental science. It is worth taking a moment to pause and think what this would mean, if true. It would mean that the vocabulary used in the special sciences – terms such as *covalent bond*, *cell wall*, *neuron*, *memory*, and so on, could ultimately be replaced by ones of a unified science. The idea could be understood as semantically *eliminative*, in that all this higher-order vocabulary could be eliminated in favour of the language of, say, fundamental physics. But even if we *could* eliminate all higher-level language, it is a rather radical claim to say that we *would* do so. There are individual cases where this has happened, though. To provide just one example, the now thoroughly abandoned idea of *vitalism* suggested that there is something fundamentally different about living organisms – namely, they are non-mechanical, containing an 'élan vital' or vital force that gives life to inanimate material bodies. The theory of vitalism involved a sophisticated theory, but the notions employed by that theory, such as that of a vital force, have been completely eliminated by contemporary science. By the late nineteenth century, the experimental progress of developmental biology and emerging modern biochemistry, combined with the lack of any experimental support for vitalism, made it clear that vitalism has no future (see Gatherer 2010 for a short history of vitalism). However, it should be noted that a plausible understanding of the unity of language should not entail full elimination of all higher-order vocabulary. Even if it were possible in principle to eliminate such vocabulary, it would surely be difficult – for instance, we do still talk about 'life' in a rather abstract fashion. Hence, there is room for non-eliminative semantic disunity.

Unity of laws, as defined by Oppenheim and Putnam, is a stronger thesis than unity of language. The thought is that all the scientific laws could be reduced to the laws of one unified science. On the face of it, unity of laws seems to be a more interesting thesis, as it does not focus on the individual terms used in science (albeit Oppenheim and Putnam do take it to imply unity of language). The precise meaning of 'reduction' is left open by Oppenheim and Putnam and I will do the same for the time being, but it appears that here we are moving towards an ontological model of unity. I will shortly discuss an example, but it may be helpful to do so in the context of the strongest sense of unity of science that Oppenheim and Putnam entertain, as it explicitly mentions also the connections within a discipline: 'Unity of Science in the strongest sense is realized if the laws of science are not only reduced to the laws of some one discipline, but the laws of that discipline are in some intuitive sense "unified" or "connected"' (Oppenheim and Putnam 1958: 4).

They specify immediately that the required 'connection' between the laws should be something stronger than the mere conjunction of these laws. What might qualify as this type of unification? A plausible candidate is provided by *electroweak unification* – the successful effort to unify two of the fundamental forces (and hence the laws concerning them), the *weak force* and the *electromagnetic force*. In fact, the electromagnetic force itself already unifies two apparently distinct forces – namely, the electric force between charges which is governed by *Coulomb's law* and the magnetic force. The *Lorentz force law* summarises both of these forces.

The unification of the weak and electromagnetic interaction is rather more complicated, as it involves further exchange particles, namely the W and Z bosons that are involved in weak interaction. Indeed, it was the prediction and discovery of the W and Z particles and the resulting unification of the weak and electromagnetic interaction that led to the 1979 Nobel Prize in physics being awarded to Weinberg, Salam, and Glashow. The press release announcing the award serves to highlight just how deeply ingrained the search for unity in science is:

> Physics, like other sciences, aspires to find common causes for apparently unrelated natural or experimental observations. A classical example is the force of gravitation introduced by Newton to explain such disparate phenomena as the apple falling to the ground and the moon moving around the earth.
>
> Another example occurred in the 19th century when it was realized, mainly through the work of Oersted in Denmark and Faraday in England, that electricity and magnetism are closely related, and are really different aspects of the electromagnetic force or interaction between charges. The final synthesis was presented in the 1860's by Maxwell in England. His work predicted the existence of electromagnetic waves and interpreted light as an electromagnetic wave phenomenon. . . .
>
> An important consequence of the theory is that the weak interaction is carried by particles having some properties in common – with the photon, which carries the electromagnetic interaction between charged particles.

> These so-called weak vector bosons differ from the massless photon primar-
> ily by having a large mass; this corresponds to the short range of the weak
> interaction.[3]

As the press release makes clear, a key part of the unification of the weak and electromagnetic forces are the shared properties of the exchange particles. The next step is to explain the primary difference, which has to do with mass. The story continues with *electroweak symmetry breaking*, and the more recent discovery of the *Higgs boson*, but we need not enter these complications. What should already be clear is that the search for unity is one of the key values of scientific inquiry, and indeed of the Nobel Committee, and there are many celebrated examples of this in the history of science. No wonder, then, that Oppenheim and Putnam attempted to systematise this idea.

We have so far omitted one important detail: when we speak of unification in science or discuss cases such as the unification of the weak and electromagnetic forces, it is not obvious that we are talking about 'unity' in the same sense that Oppenheim and Putnam or contemporary philosophers always intend. There is clearly something in common with these cases, which is why I have used electroweak unification as an example of how the laws of physics might be 'connected' in the sense that Oppenheim and Putnam require. But we should keep in mind the distinction between ontological models of unity and epistemic/pragmatic models of unity. In particular, the case I have just described could be understood in the sense of theoretical unity (or unity of formalism), so we should consider this form of unity in more detail.

The idea behind theoretical unity is simply that we may discover a formal (mathematical) framework, which manages to approximately model a certain set of distinct phenomena. For a given purpose, it may be sufficient to use a simple unified formalism. To continue our previous example, consider the role of the electromagnetic force in holding together atoms and molecules. The electromagnetic force is by far the most significant force in determining atomic and molecular structure. It has an infinite range, just like gravity, but given the extremely small masses of particles in the atomic scale, gravity is negligible. The strong force, by contrast, is very strong indeed, but its range is very short – it holds the nucleus together. The weak force has an even shorter range, 0.1% of the diameter of a proton. If we are interested in the molecular range, it is really just the electromagnetic force that matters. So, for most calculations that we might wish to make concerning the molecular scale, it is entirely unnecessary to consider gravity, *even though* gravity is in effect at all scales. Thus theoretical

---

[3] Press release, NobelPrize.org, accessed 27 June 2019. www.nobelprize.org/prizes/physics/1979/press-release.

unity has an element of interest relativeness, which is useful and even necessary for science, but it is not the type of ontological unity that some philosophers may be interested in.[4]

In contrast, ontological models of unity concern the structure of reality rather than the structure of theories. Yet, so far, we have not specified a properly *ontological* as opposed to *epistemological* sense of reduction that might underlie such unity. Let us turn back to Oppenheim and Putnam, who also state that they wish to set aside epistemological versions of the unity thesis (cf. Oppenheim and Putnam 1958: 5). Relying on previous work by Kemeny and Oppenheim (1956), they take reduction to be a relation between *theories*. This may not quite capture the sense of ontological unity that I have just been alluding to, but we should not be misled by this, for Oppenheim and Putnam do specify that a key part of reduction is that a set of observational data *explainable* by one theory is explainable by the reducing theory. This explanatory connection, we may assume, is supposed to track the ontological relation between the phenomena that the theories describe. What is that relation? Oppenheim and Putnam call it *micro-reduction*. As they specify, this relation concerns the *objects* or entities that theories deal with, so it is ontological rather than epistemological in the intended sense.

The idea of micro-reduction is something that survives in contemporary philosophy (under different labels), so it is useful to consider it in some detail (cf. *microstructural essentialism*, which we will return to later; see also Tahko 2015b). Micro-reduction is transitive, irreflexive, and asymmetric. As Putnam and Oppenheim (1958: 7) observe, the transitivity of the relation is of particular importance since it establishes a hierarchy of *reductive levels*. The thought is that there must be more than one such level (rather than a 'flat' one-level reality), there is a unique lowest level (such as fundamental physics), and a common denominator for each level (any thing on one level, except for the fundamental one, must be composed of parts on the level immediately below it). In practice, this means that if psychology reduces to neuroscience and neuroscience to biochemistry, then in virtue of transitivity, psychology will also reduce to biochemistry. This strict hierarchical structure may seem controversial because sometimes it does seem that we have to consider two levels at more extreme ends. For instance, the field of *quantum biology* applies results from

---

[4] Indeed, it is ontological unity that I am most interested in, instead of theoretical unity, which is primarily epistemic. For a thorough discussion of theoretical unity (in physics), see Morrison (2000). It is worth keeping this distinction in mind, because the pursuit of theoretical unity has such an important role in science. It may often also point to ontological unity, but it does not entail it. Note also that one may of course be interested in *both* theoretical and ontological unity, as many philosophers of science surely are.

quantum mechanics directly to the realm of biological entities. But Oppenheim and Putnam insist that we better not 'skip' any levels in this process, say, by trying to explain psychology in terms of subatomic physics, as this would indeed be 'fantastic' (ibid.). We now arrive at Oppenheim and Putnam's proposed account of the levels of science, represented by the pyramid in Figure 2.

Notice that a key idea driving reduction in Figure 2 seems to be a simple parthood or composition relation: social groups are composed of living things, which are composed of cells, which are composed of molecules, and so on. This is a type of 'building blocks' conception of nature that has survived a long time in philosophy and still enjoys support, e.g., in the form of a commitment to a fundamental compositional level (see Tahko 2018 for discussion). Note, however, that a commitment to a fundamental level does not by itself entail such a conception – for one thing, there could be just one level, in which case the 'building blocks' idea would not be a good fit. Moreover, if the various levels of entities in Figure 2 are thought to be genuinely existing, then the model at hand would be compatible with non-reductive ontological disunity, while eliminating the higher-level entities in favour of a singular level of elementary particles would entail reductive ontological unity.

The result, then, is a hierarchical picture of the levels of reality which is really driven by just *one* relation – namely, the parthood or composition relation. This is effectively what the idea of micro-reduction amounts to on the Oppenheim-Putnam line. However, it is worth keeping in mind that Oppenheim and Putnam did not claim (and presumably would not claim even now, more than sixty years later) that we already possess the tools to perform this type of micro-reduction. Rather, '[t]he assumption that unitary science can be attained through cumulative micro-reduction recommends

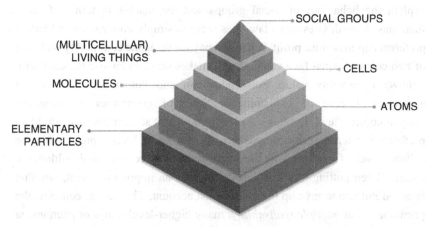

Figure 2 A system of reductive levels (author's own work)

itself as a *working hypothesis*' (Oppenheim and Putnam 1958: 8). By a 'working hypothesis', they mean that this idea should be justified on empirical grounds, if and when we are able to establish this type of reduction. They go on to support this idea (1958: 12ff.) in terms of theoretical virtues such as the simplicity of the hypothesis and the variety and reliability of the evidence in favour of it from the various sciences and successful inter-theoretic reductions. They also put forward case studies from each of the levels. Among the more famous case studies, partly because it was later picked up by Fodor, is the case of Gresham's law in economics.

Gresham's law, which states that bad money drives out good, can be illustrated with a simple example. In this example, 'good' money is money, such as gold coins, where the nominal value (the face value) of the coin is equal or close to equal to the value of the metal it is made of (the commodity value), in this case gold. 'Bad' money, on the other hand, is money which has a higher nominal value than commodity value. So, Gresham's law states that when 'bad' money is circulating along with 'good' money of the same nominal value, it tends to drive out the 'good' money, since the higher commodity value of the 'good' money is an incentive to hold on to it. A consequence of this law was the practice of debasement, which could unlawfully be done by the public – for example, by scraping off small portions of gold from gold coins, while the coin retains its nominal value. The reeded edges on coins, still present in many modern-day coins (such as the two-euro coin), were intended to make this practice evident and hence allow the recipient to deny a payment in such 'bad' money.

Oppenheim and Putnam suggest that we can analyse cases such as Gresham's law in terms of individual choices, where these choices are ordered by means of an 'individual preference function' (1958: 17). Economists can then attempt to explain the behaviour of social groups and the market in terms of these functions. Indeed, Gresham's law does seem to simply confirm the individual preference to maximise profit: if you can pay for the same commodity with one of two coins of equal face value, then it makes sense to choose the coin with a lower commodity value. So, Oppenheim and Putnam take this to be a successful case of explaining a 'law' of economics in terms of a psychological 'law'. But it is less clear how we can account for this individual preference in terms of its composite parts at a lower level – presumably the cellular level of neurons – as the micro-reductive account would ultimately require. Even putting this worry aside, there is an important caveat, one that later led Putnam to give up the reductionist account. This caveat concerns the phenomenon of *multiple realisability*: many higher-level things or phenomena can be realised by two or more lower-level things.

2.3 Fodor on Multiple Realisability and the Disunity of Science

Jerry Fodor, in two very influential papers (1974, 1997), argues against a strong form of reductionism and in favour of the autonomy of the special sciences. The first of these papers reacts against the traditional Nagelian model of reaction described in Section 2.1, while the second addresses a rejoinder to Fodor's original paper by Jaegwon Kim (1992). There are two important methodological points to note about this line of discussion that began in the seventies.[5] First, very much unlike the logical empiricist tradition or Oppenheim and Putnam's traditional discussion, this new literature was concerned with *metaphysics* of science (even though this notion was not used): the focus is on notions of natural kind, property, realisation, and so on. These are metaphysical notions that rarely feature directly in scientific theories, and we will discuss their role in more detail later on. Second, despite the import from metaphysics of science, there is in fact a much closer interest to actual scientific case studies – a trend that we see continuing in contemporary metaphysics of science. The latter methodological point is especially striking, given the supposed scientific rigour of the logical empiricist tradition.

Fodor, in fact, starts by noting that the trend toward reductionism in philosophy of science (at that point in the seventies), while driven by scientific successes, is not thoroughly explained by it. Instead, Fodor thinks that many proponents of reductionism are motivated by the relative *generality* of physics when compared to the special sciences. But he thinks that this idea of the generality of physics should be separated from a stronger sense of 'the unity of science':

> What has traditionally been called 'the unity of science' is a much stronger, and much less plausible, thesis than the generality of physics. If this is true it is important. Though reductionism is an empirical doctrine, it is intended to play a regulative role in scientific practice. Reducibility to physics is taken to be a *constraint* upon the acceptability of theories in the special sciences, with the curious consequence that the more the special sciences succeed, the more they ought to disappear. (Fodor 1974: 97)

The thought that the special sciences ought to disappear as they succeed is indeed a rather strong doctrine. It reflects the strong reductionism of logical empiricism and it may or may not be the case that Oppenheim and Putnam were committed to something as strong as this, but such a view is certainly not very popular today, even among those who would call themselves reductionists (Rosenberg 1994, e.g., may be an exception). So, Fodor's target is a very strong

---

[5] Thanks to Carl Gillett for highlighting these issues.

sense of unity, and so is his understanding of reduction. He characterises it in terms of bridge laws, where predicates of the reducing and reduced science are connected via an identity relation – for example, every event which consists of x's satisfying some neurological predicate S_1 is identical to some event which consists of x's satisfying some biochemical predicate S_2 and vice versa (cf. Fodor 1974: 100). The resulting picture looks like a combination of reductive ontological unity and eliminative semantic unity.

Let us return to Gresham's law. One of Fodor's central points is that not every natural kind corresponds to a *physical* natural kind. The notion of a 'natural kind' is here interpreted very liberally: it is a general term that captures something that is in common among the instances of a kind. We will return to the interpretation of natural kinds in much more detail in Section 3. So, monetary exchanges, which are what Gresham's law concerns, could be regarded as a natural kind insofar as there is something in common among its various instances. Fodor (ibid., 103; original emphasis) states the following: 'I am willing to believe that physics is general *in the sense that it implies that any event which consists of a monetary exchange* (hence any event which falls under Gresham's law) *has a true description in the vocabulary of physics and in virtue of which it falls under the laws of physics.*' But Fodor then points out that since monetary exchanges are realised by a vast variety of different realisers (not just euro coins but also dollar bills, cheques, and 'strings of wampum'), the events being covered by the reducing science would have to be 'wildly disjunctive'. The worry is that any disjunctive physical predicate that would be able to cover all these instances of monetary exchanges would not express a physical natural kind (and would not feature in laws of physics). Yet monetary exchanges presumably have something interesting in common despite the variety of different realisers. So, Fodor's point is that there are other, higher-level natural kinds in addition to physical kinds: 'A natural kind like a monetary exchange *could* turn out to be co-extensive with a physical natural kind; but if it did, that would be an accident on a cosmic scale' (Fodor 1974: 104). The upshot, according to Fodor, is that economics is not reducible to physics and hence there is no unity of science.

We have now arrived at a key turning point in the history of reductionism and indeed the search for a unity of science. Even though Fodor did not yet use the notion of *multiple realisation* in his 1974 paper, this is clearly the idea behind the 'wildly disjunctive' realisers of monetary exchanges. But in fact, Putnam had already discussed the idea of multiple realisation in the late sixties (Putnam 1967). Fodor (1974: 105) mentions Putnam's ideas, but does not cite his work (indeed, he cites only Chomsky besides himself!). A little later in the paper (ibid., 108), he specifies that there is 'an open empirical possibility' that the realisers of a supposedly reducible higher-level natural kind predicate could turn out to be 'a heterogeneous and unsystematic disjunction of predicates in the reducing science, and we do not want the unity of science to be prejudiced by this possibility'. Fodor's focus on *predicates* is problematic though, for why should we think that any (or all) predicates of the special sciences should correspond to genuine natural kinds? Indeed, we should not think so, any more than we should think that Nelson Goodman's famous predicate 'grue', which applies to all things examined before some future time *t* if and only if they are green but to other things observed at *t* or after it if and only if they are blue.

Of course, 'grue' is a perfectly legitimate predicate, and we can imagine that such a made-up predicate could have some use in a specific scenario. Yet there is no reason whatsoever to suppose that there should be a physical natural kind predicate that corresponds to 'grue'. As we shall see, it is thus better to focus on properties rather than predicates, in order to avoid the clearly conventional elements associated with predicates. I take this to be a serious flaw in Fodor's original criticism of reductionism. He concludes his 1974 criticism of strong reductionism by saying that it is not 'required that taxonomies which the special sciences employ must themselves reduce to the taxonomy of physics' (ibid., 114). But this is something that a friend of reductive ontological unity can agree with: ontological reduction does not concern the *taxonomy* of the special sciences. In other words, reductive ontological unity is compatible with non-eliminative semantic disunity. Admittedly, the strong version of reductionism originally pushed by the logical empiricists in the form of eliminative semantic unity does entail eliminativism regarding higher-level taxonomy, and this is Fodor's main target, but it would be a mistake to regard the unity of science quite generally to be hostage to the unity of taxonomy.

2.4 Kim versus Fodor on Jade

Moving on to the next stage of the debate, we should take a closer look at Jaegwon Kim's famous discussion of multiple realisation concerning the case of jade, along with Fodor's reply to this case. This example was originally introduced by Putnam (1975: 241), although for a slightly different purpose. Kim summarises the case as follows: '[W]e are told that jade, as it turns out, is not a mineral kind, contrary to what was once believed; rather, jade is comprised of two distinct minerals with dissimilar molecular structures, *jadeite* and *nephrite*' (1992: 11). He then introduces a supposed special science law stating that jade is green and divides that into a conjunctive law, whereby jadeite is green and nephrite is green. Here, 'jade' is of course supposed to refer to a higher-level natural kind. But Kim, favouring reductionism, thinks that the special science law in question is not a genuine law, despite having the basic form of a law and being able to support counterfactuals. This is because the law is not *projectible* – it does not have 'the ability to be confirmed by observation of positive instances' (ibid.). Kim's case in favour of this conclusion is the possibility that we only encounter one or the other of the realisers of jade when trying to confirm the law stating that jade is green. This would only support the projectibility of either jadeite or nephrite being green. The upshot, according to Kim, is that jade is a 'true disjunctive kind' (Kim 1992: 12).

Now we get to Kim's famous challenge for Fodor: he argues that special science laws concerning 'wildly disjunctive' higher-level kinds are not genuine laws, because they do not deal with genuine natural kinds. They can be reduced to laws concerning lower-level kinds. But Fodor's reply to this challenge is also forceful; he denies that jade is paradigmatically multiply realisable in the first place and, further, claims that the scenario Kim describes is just based on a sampling error. However, Fodor does admit that jade is a true disjunctive kind; what he denies is that this is the same thing as being *disjunctively realised*: 'Jade is disjunctive because the only metaphysically possible worlds for jade are the ones which contain either [jadeite], or [nephrite] or both. By contrast, multiply based properties that are disjunctively realized have different bases in different worlds. Pain is disjunctively *realized* because there's a metaphysically possible, nonactual, world in which there are silicon based pains' (Fodor 1997: 153).

For Fodor, the case of jade is not a genuine case of disjunctive realisation, so he agrees with Kim that it is not projectible in the required sense. Moreover, the disjunction is 'closed' in that in all cases we are dealing with either jadeite or nephrite, and there are no further candidates. This constitutes a case in favour of abandoning the special science law concerning jade and replacing it with lower-level laws concerning jadeite and nephrite. But if we are dealing with an 'open' disjunction, then there are metaphysically possible worlds in which a property has realisers that it does not have in the actual world – only this type of case counts as genuine multiple realisation for Fodor (Fodor 1997: 156). So, the corresponding lower-level laws would also have to be open, and Fodor insists that 'if there is a higher level property that subsumes all the disjuncts of an open disjunction, then we will want to state our laws in terms of it' (ibid., 158).

The idea behind Fodor's open/closed distinction is, I take it, to postulate higher-level kinds in order to get the desirable closed laws back, *unless* we can point to closed lower-level laws. Fodor's claim, then, is that while the case of jade is closed, the case of pain (or other genuinely disjunctively realised *functional kinds* – kinds which are defined in terms of what they do) may be open and hence justify postulating higher-level kinds. Fodor summarises this idea with the following principle that he takes to govern our inductive practises: 'Prefer the strongest claim compatible with the evidence, all else equal. Quantification over instances is one aspect of rational compliance with this injunction; reification of high level kinds is another' (Fodor 1997: 159).

One thing that is striking about this debate is just how little scientific detail it involves, especially when the key argument is supposed to be that it is *science* that gives us the criteria according to which we ought to judge higher-level kinds. This is particularly ironic since Kim and Fodor supposedly agree about the case of jade. What they do not seem to realise is that jadeite and nephrite, the supposed realisers of jade, are *themselves* higher-level kinds. In fact, Fodor (1997: 154) makes a passing appeal to the commonly accepted Kripke-Putnam framework of *microstructural essentialism*, stating that jade's being jadeite or nephrite is metaphysically necessary, just like it is metaphysically necessary that water is H_2O. According to this framework, kinds are defined in terms of their microstructural essence. That's how Fodor arrives at the judgement that jade is not a genuinely disjunctively realised kind. But this goes against Kim (1992: 24), who suggests that jade is to be defined in terms of its *macrophysical* properties.

Fodor may think that his modal intuitions 'are pretty clearly the right ones to have' (Fodor 1997), but it has been clear at least since Jaap van Brakel's (1986) work (see also VandeWall 2007) – over a decade before Fodor's snarky remark – that even the usual microstructural story about water is seriously wanting.[6] In particular, to arrive at the result that the case of jade is closed in Fodor's sense, we must rely on the controversial assumption that the macrophysical properties of jade are irrelevant. This is the upshot of Fodor's speculative scenario where someone recreates the macrophysical properties of jade from melted bottle glass. Fodor insists (based on the 'pretty clearly right' modal intuitions) that the result would not be jade. But this already assumes that the higher-level kind jade is genuine and to be defined in terms of the traditional microstructuralist story. Rather than take this assumption for granted, we would do well to consider what we know about jadeite and nephrite.

We know that jadeite and nephrite share many of their chemical properties yet differ in terms of microstructure. Jadeite and nephrite are not exactly identical in terms of their chemical properties. Jadeite ($NaAlSi_2O_6$) is somewhat harder and less prone to scratches due to its dense crystal structure and higher specific gravity – it is a pyroxene mineral. Nephrite ($Ca_2(Mg,Fe)_5Si_8O_{22}(OH)_2$) is a mineral in the actinolite-tremolite series. I would like to draw attention to

---

[6] For a more recent, thorough discussion of the case of water and microstructuralism more generally, see Needham (2011) and Tahko (2015b).

one aspect of nephrite's chemical formula: (Mg,Fe). This peculiar feature means that different samples of nephrite can have varying proportions of Mg and Fe. This is not just a feature of nephrite. Consider the common mineral *olivine*, which also occurs in two varieties, a magnesium-rich and an iron-rich variety; this is similarly reflected in its chemical formula, $(Mg,Fe)_2SiO_4$. The chemical properties of olivine vary according to whether it is Mg-rich or Fe-rich (e.g., only the latter can exist stably with silica minerals such as quartz). So, are olivine and nephrite disjunctive kinds like jade in that they are realised, in all possible worlds, either by the Mg-rich or the Fe-rich variety? Or are they genuinely disjunctively realised, like Fodor takes pain to be?

In fact, neither answer seems correct. The reason is that minerals such as olivine (and, more generally, feldspars and pyroxenes) are usually considered as *mixtures* rather than compounds; they are more appropriately understood as *solid solutions*.[7] A solution is a form of mixture, but in this case, the solution is of course much more rigid than the solutions we usually encounter. The principle, however, is very similar. Think of a vinaigrette emulsion, which is also a type of mixture: the olive oil and the vinegar are normally immiscible, which means that they do not naturally form a homogeneous mixture. But by shaking them into an emulsion, we have created a homogeneous mixture. This is a simple example of the type of mechanism that we have in the case of solid solutions as well.

The upshot is that we should define jadeite and nephrite as distinct kinds (if they are to be understood as kinds at all), which, as Kim suggests, have been classified as jade due to some superficial, macrophysical similarities rather than some shared microstructural basis. Hence, if we follow Fodor's microstructuralist criteria, there are good, scientific reasons to think that these minerals are not genuine natural kinds at all; they are mixtures of two elements in close proximity on the periodic table that remain in a stable, homogeneous state, e.g., when combined with silica minerals (compare this with VandeWall 2007 on the case of H_2O). This does not mean that the general microstructuralist approach that Fodor relies on must fail, but in the case of the particular example being discussed, Kim's macrophysical approach is evidently much better supported by scientific practice. Insofar as we wish to postulate higher-level kinds such as jade, we ought to define them in terms of their macrophysical properties. In scientific practice, jadeite and nephrite are clearly distinguished, since they differ in terms of their chemical and many of their macrophysical properties, but note that it may nevertheless be possible to adopt a theory that allows

---

[7] For a more technical account, see, for instance, Nesse (2011) for an introduction to mineralogy and solid solutions.

genuine natural kinds to be macrophysically defined (cf. Hendry 2010, Needham 2011, and Tahko 2015b; see also Seifert 2019).

It seems then that the case of jade certainly does not settle the debate between reductionism and anti-reductionism – in that, Fodor is correct. But Kim's analysis of the case raises legitimate concerns about the anti-reductionist approach. Let's see if we can put our finger on these concerns independently of the case of jade.

2.5 Are Higher-Level Kinds 'Really There'?

Let us move on to an interesting analysis of the Kim-Fodor debate, by Louise Antony:

> [M]ultiple realizability is something of a red herring. What matters, funda-mentally, is not whether there could be minds embodied in things other than brains, but rather whether there is a level of reality beyond the level at which brains are normally studied – whether psychological kinds are 'really there,' 'over and above' the already recognized kinds in chemistry, biology and the other established sciences. (Antony 2003: 8)

The crucial question is: what makes a kind 'real' and can disjunctive kinds indeed be 'real' in the relevant sense? Antony draws on a suggestion from Lenny Clapp (2001), arguing that disjunctive properties can indeed be 'real' in cases where the disjuncts 'have real commonalities' (Antony 2003: 10). The question then turns entirely on whether there is such a real underlying resem-blance in the case of disjunctive predicates.

Somehow, the anti-reductionist must define an objective sense of higher-level 'reality'. But I find Antony's ultimate case for this unconvincing; she suggests, very much in line with Fodor, that there is a 'strong abductive argument from the *projectibility* of higher-level predicates to the *reality* of the kinds they designate' (ibid., 13). The idea is that the disjunctive predicates designating 'real' kinds are necessarily co-extensive with some projectible predicates (but she qualifies that this is merely a sufficient rather than a necessary criterion). Antony suggests that this is the case with many predicates 'entrenched' in human language – reflecting Nelson Goodman's view in letter even if not in spirit (since Goodman was deflationary about projectibility). However, as Antony immediately acknowledges, there is an issue that might easily under-mine such an argument: there are plenty of projectible higher-level predicates that are suitably entrenched in human language, but which we surely do not wish to reify as genuine, higher-level natural kinds. Two examples of this are 'witch' and 'angel' – predicates that we clearly do not believe to correspond with genuine natural kinds. However, both 'witch' and 'angel' may be projectible, because we can, for instance, make true generalisations about the women labelled as witches in the Middle Ages. It is just that those generalisations are not true because these women were *witches* with some supernatural skills, but rather because they were women who were all oppressed for some social or political reasons. So, what should we say about such cases?

Antony has an interesting suggestion about how to deal with this. One could try to deny that the abductive argument goes through (and hence deny the entrenchment of predicates like 'witch') in these cases, because there really have not been that many successful predictions made using these predicates. Or one could explain the entrenchment in terms of something else than what the supernatural connotations regarding 'witch' or 'angel' suggest, namely, in terms of socio-political groupings, as suggested previously. The reason why this suggestion is interesting is that it gives us a nice way to distinguish between eliminativism and reductionism:

> [T]he eliminativist about the mental denies that mentalistic predicates are projectible. Such a person disagrees with the mental realist about the robustness of our psychological attributions, denying either that there are any substantive predictions that can be made on the basis of such attributions, or else that the predictions that are so based are successful. The reductionist, on the other, agrees that psychological predicates are projectible, but thinks that they are so only because they track *biological* kinds. (Antony 2003: 14; original emphasis)

Antony's proposal nicely illuminates the eliminativism-reductionism distinction. Both are strategies to avoid including things like witches and angels among the genuine natural kinds. So, could one of these strategies apply in the case of jade?

Antony does not seem to think so. Like Kim and Fodor, she disqualifies jade from the class of 'real' natural kinds, appealing to Fodor's argument (based on the distinction between open and closed disjunctions), which we outlined in the previous section: 'It's a mere *accident* that both jadeite and nephrite count as jade; there is nothing that the two mineral kinds [jadeite and nephrite] *really* have in common' (Antony 2003: 15). Or even if there is, she continues, it is not in virtue of the observable macrophysical similarities that we count jadeite and nephrite as jade, contra Kim – this reply reflects the case of witches and angels. But I think that here's where the anti-reductionist strategy that Antony develops breaks down: 'jade' satisfies all the criteria that were given for 'real' kinds in Antony's theory. It is entrenched in human language, and it is entrenched precisely because there is a set of higher-level, macrophysical properties that we associate with 'jade' and that we value, such as aesthetics and ease of working into exquisite objects (cf. Hacking 2007a: 276; LaPorte 2004:

94–100). These properties are had both by jadeite and nephrite because, despite the differences in microstructure, both result in sufficiently (from the point of view of entrenchment) similar macrophysical properties. So, the abductive argument goes through here and it cannot be explained away by either of the strategies that Antony entertains. The reason for this is that we did not make any mistake here about what is projectible about jade – we were always interested in the macrophysical properties of jade, whereas in the case of witches we were, supposedly, interested in their dealings with the devil. Indeed, it is useful to refer to both jadeite and nephrite with the same term precisely because they share roughly the same, valued macrophysical properties, and it is these properties that secure its projectibility.

It should be getting clear what the upshot of all this is for the unity of science. Everyone agrees, we were told, that jade is not a 'real' natural kind, and hence not a legitimate counterexample to Fodor's account of the disunity of science. But the criteria for 'real' make jade every bit as real as the postulated higher-level kinds that the anti-reductionists typically wish to reify. I think that a more systematic account of what it takes for something to be a genuine natural kind is needed. If it turns out that there really are genuine, irreducible higher-level natural kinds, then the traditional pluralist argument for the disunity of science is back in business. But if we can unify our theory of natural kinds and either reduce or explain away the supposedly irreducible kinds, then higher-level kinds do not threaten reductive ontological unity. Fortunately, there has been plenty of progress in our theory of natural kinds in recent years. In Section 4, we will return to the issue of natural kinds and its connection to unity. But first, we should discuss the state of the art regarding unity of science.

3 Combining Unity and Pluralism

We have now seen that there are many different ways to interpret the unity of science: in terms of the various theories, models or formal frameworks used in science, in terms of laws and kinds postulated in science, and in terms of dependence relations like composition that connect the entities studied in science. An important distinction introduced in Section 2 in this regard is the one between ontological and epistemic/pragmatic models of unity (the reader is invited to refer to Figure 1). We have also seen that the phenomenon of multiple realisability led to the popularity of different forms of pluralism, whereas strong forms of reductionism have been largely abandoned. Moreover, where reductionism is still pursued, there has been a shift of focus from *global* reductive projects towards *local* reduction. We can also see an active focus on attempts to clarify the notions of reduction, multiple realisability, and emergence (e.g., van

Riel 2014, Wilson 2015, forthcoming, Gillett 2016, Polger and Shapiro 2016, and many others). Each of these topics is worthy of books longer than this Element, so I will not attempt to do justice to all the nuances in this literature. We have discussed reduction and multiple realisation in some detail already, but we should briefly note the connection to emergence as well.

Emergence can very generally be described as the combination of dependence with autonomy: something is emergent if it depends on something else but nevertheless maintains some important sense of autonomy. The literature is full of debate about what constitutes the relevant senses of dependence and especially autonomy. In the present context, a potentially more interesting feature is the relationship between reduction and emergence. In particular, emergence is typically considered to contrast with reductionism – one possible sense of 'autonomy' is simply 'non-reducible'. Yet it is certainly possible to combine weak forms of reductionism and weak forms of emergence. A helpful distinction can be made between *weak emergence* and *strong emergence* (e.g., Wilson 2015). Weak emergence in its various forms is quite widely accepted, since it generally requires some modest form of autonomy for the emergent (higher-level) phenomenon; this could be as simple as the emergent phenomenon being unexpected or independent with regard to some minor changes in its dependence-base (lower-level phenomena). An example of weak emergence might be the various surprising products of evolution (such as intelligence). We might not fully understand or expect to see all these products, but we know that they are the result of complex biological and chemical processes. Strong emergence is much more controversial, as it requires autonomy to a very high degree – for example, it is common to think that a strongly emergent phenomenon is not even in principle deducible from lower-level phenomena. It is unclear whether strong emergence exists, but consciousness is often suggested as a potential candidate.

This very brief introduction to emergence is not supposed to be comprehensive, but I hope that it suffices to demonstrate why someone interested in unity of science might also be interested in emergence – for strong emergence would seem to immediately challenge at least the possibility of stronger forms of unity, since it prevents strict reduction and arguably also weaker forms of reduction. Moreover, the autonomy of emergent phenomena, even when the emergence is weak, does call for further elucidation. Importantly, there may be ways of understanding that autonomy that pose challenges for reductionism. Accordingly, there is much to explore regarding the interplay of these various notions. I will make some further remarks about this interplay where relevant in what follows, but a thorough discussion of emergence will have to take place elsewhere.

In any case, even if some higher-level phenomena are emergent, there is also a sense of unity that we need to account for. One pressing issue here is precisely the sense in which unity of science may or may not be compatible with the autonomy of the special sciences. In other words, can we combine a sense of unity and pluralism? We can see this theme already in the discussion between Fodor, Kim, Antony, and others, where the role of multiple realisability is central. It could be said that the attempt to accommodate the phenomenon of multiple realisability while maintaining a form of ontological reductionism is one way to combine unity and pluralism. But in more recent literature, we can find sophisticated attempts to adopt a sense of pluralism without necessarily focusing on multiple realisability. I will start by tackling this issue, before presenting two case studies where we see the combination of unity and pluralism in action. The notion of 'pluralism' could mean a number of things, such as a plurality of composed and component entities and kinds or a plurality of higher-level explanations and predicates. These two broad senses of pluralism roughly correspond with the distinction between ontological and epistemic/pragmatic models of unity. In what follows I will understand 'pluralism' in the second sense (i.e., as a type of semantic pluralism about higher-level talk). However, we will return to ontological forms of pluralism in Section 4.

There is a sense in which a modest interpretation of unity of science is obviously true. This sense is appropriately captured by Sandra Mitchell (2002: 55), who asks, '[I]f science is representing and explaining the structure of the *one* world, why is there such a diversity of representations and explanations in some domains?' So, insofar as the world is, in some sense, a unified whole, why do our scientific methods not follow suit? This is an important question, but it should be immediately noted that it is primarily a question about scientific *practice* and our epistemic limitations, and hence in the realm of epistemic/pragmatic models of unity. One reaction to the question, which Mitchell herself considers, is to say that the variety of different methods and models in science is a sign of *incomplete* science. That is, since we do not yet have a full picture of the world, it is understandable that science approaches the world in a piecemeal fashion, sometimes with incompatible methods and models, while ultimately pursuing theoretical unity. This would make pluralism a purely epistemic issue. But even though science certainly sometimes proceeds towards theoretical unity, this is not always the case even if an integrated model were available. The reason for this, as Mitchell notes, is the *complexity* of the subject matter. It is simply not effective to strive for theoretical (or methodological) unity across the sciences, because scientific research would be less effective without a plurality of approaches. Now, this does not mean that unity of science has been lost; rather, it means that pluralism thus conceived is an epistemic or pragmatic issue. We can see elements of this type of pluralism in the work of Nancy Cartwright (e.g., 1999) and William Wimsatt (1987) as well, but Mitchell's *integrative pluralism* goes somewhat further in endorsing *competing*, idealised models in science. However, the resulting pluralism of models does nothing to undermine ontological models of unity (nor is it intended to do so): as Mitchell concludes, at the theoretical level, pluralism is allowed (even required), but at the 'concrete explanatory level' (i.e., when the models are interpreted and applied to the real world), integration is required (Mitchell 2002: 66).[8]

I believe that something in the lines of Mitchell's quite sensible approach to pluralism is now a widely shared view, although there are exceptions – those who think that a more radical, metaphysical pluralism is required or who favour eliminative semantic unity. But how does this type of epistemic/pragmatic pluralism compare with ontological models of unity? What is the underlying metaphysical picture here? One thing that is striking is that even though the possibility of combining theoretical or methodological pluralism and onto-logical models of unity does seem to be widely accepted, there have been very few attempts to clarify the underlying metaphysics, Carl Gillett's work being one notable exception. Perhaps part of the reason for this is that this metaphysical picture generally has reductionist elements, and reductionism, for many, still has an unfavourable ring to it (for a classic example, see Dupré 1983). But once we acknowledge that even reductive ontological unity does not entail semantic eliminativism, and is hence fully compatible with *non-eliminative* semantic disunity and theoretical pluralism, there is no reason to associate it with eliminative semantic unity of the form adopted by the logical empiricists.

To see all this, it is finally time to specify the type of reductionism that we may associate with reductive ontological unity. This type of *ontological reductionism* is a relatively weak form of reductionism (cf. also Oppenheim and Putnam's 'micro-reduction', discussed in Section 2.2):

Ontological reductionism (OR): The view that true statements about higher-level phenomena are made true fully and only by lower-level phenomena.

---

[8] Compare this also with Breitenbach and Choi (2017), who argue that despite the plurality of methods in science, a commitment to the unity of science as a regulative ideal is useful. Their account is explicitly epistemic – in other words, Breitenbach and Choi consider ideal unified science to be a useful epistemic aim, even if it were unreachable in practice. Other recent primarily epistemic accounts of unity can be found in Nathan (2017) and Patrick (2018), but the best-known account is probably Philip Kitcher's, going back at least to Kitcher (1981), so this is not exactly a new trend.

Here, the relevant phenomena may be understood, for example, as properties, kinds, powers, individuals, or mechanisms. The thought is that, for instance, statements about a higher-level entity that is composed of lower-level entities are made true by those component entities. It is important to see that OR does *not* suggest that we should replace our statements about higher-level entities with statements about lower-level entities. Hence, OR does not entail eliminative semantic unity, it is not eliminative about higher-level talk. This point about the compatibility of ontological reductionism and anti-eliminativism has probably not been fully appreciated, even though it has been made, sometimes implicitly and sometimes explicitly, by several authors in recent work (e.g., Heil 2003a, Gillett 2007, Wilson 2010, and Strevens 2012).

For instance, as Gillett outlines the view he calls 'new reductionism', our focus should be on the ontological relations between the relevant entities that the sciences investigate (again, the relevant entities may be properties, kinds, powers, individuals, mechanisms, or something else, see the articles in Aizawa and Gillett 2016 for some examples). Gillett explicitly states that his own version of this, dubbed *compositional reductionism*, should be combined with *semantic anti-reductionism* about 'special sciences and their predicates' (Gillett 2007: 195). Semantic anti-reductionism is the view that we can maintain the higher-level kind predicates and special science practices, while at the same time acknowledging that there is a lower-level story that we can tell about those practices. Gillett's compositional reductionism is one way to tell this story, as it 'takes our initial ontological commitments to both component entities and composed higher-level powers, properties, individuals and processes, and reduces them to a commitment solely to *one* layer of truthmakers in the relevant component entities' (ibid.). The resulting picture *is* eliminative in another sense though, as it eliminates a commitment to higher-level entities as truthmakers for higher-level talk. So, while we may postulate compositional and other relations between higher-level and lower-level entities, strictly speaking, these do not exist because the higher-level entities do not exist. However, I will reserve the term 'eliminativism' for *semantic* eliminativism and talk about ontological reductionism and reductive ontological unity to capture this ontological sense of eliminativism.

Michael Strevens makes a related point in terms of explanation: '[E]xplanatory physicalism is perfectly compatible with the view that the explanation of certain phenomena is best conducted at a rather abstract level, omitting those details of physical implementation that make no difference to the phenomena's obtaining' (Strevens 2012: 755). Here, 'explanatory physicalism' is the view that everything that can be explained can be explained physically. Strevens then goes on to argue that the abstract level of explanation of the higher-level sciences involves properties that are *irreducible*. But as a matter of fact, this is a little underspecified, because Strevens thinks that a property is irreducible 'just in case it cannot be defined in physical terms' – by using physical *vocabulary* (Strevens 2012: 755–756). So, the point seems to be very much in line with what Gillett calls semantic anti-reductionism: we need the predicates of higher-level sciences for our epistemic, explanatory purposes. In other words, the focus is really on higher-level predicates rather than, say, higher-level natural kinds. Contrary to what Strevens (ibid.) somewhat misleadingly claims, this is not a sense of irreducibility that is 'very strong'. In fact, it is just about the weakest form of irreducibility that there is, as it only requires that some higher-level predicates are indispensable for our explanatory needs, where 'explanatory' is to be understood as an epistemic rather than a metaphysical notion. Why would anyone deny this?

Strevens does make one very interesting additional contribution, which is a significant improvement over the traditional debate concerning reductionism. This is his point about the need, or lack thereof, for *bridge principles* (or bridge laws), which would provide necessary and sufficient conditions enabling a reductive connection between two levels. Already in our discussion of the debate between Fodor and Kim, we saw that this type of link is likely to be too strong a requirement, reminiscent of Nagelian reduction. Clearly, such bridge principles are rarely available – a point that Strevens bolsters.

I should mention that the physicalism that Strevens has in mind here is of course of the 'non-reductive' sort: even if some higher-level kinds are irreducible, they are still physically realised (cf. Strevens 2012: 756). In other words, every instance of money is physical (even bitcoins), even if the kind *money* is, in some sense, not. This is apt to cause some terminological confusion, but the distinction is a familiar one: the 'non-reductive' or 'token' physicalist view is contrasted with fully reductive 'type' physicalism. The same is true for Jessica Wilson's version; she outlines and argues in favour of weak emergence, which is enabled by the elimination of some degrees of freedom regarding the weakly emergent entity's realisation base (for details, see Wilson 2010: 292). The more general idea underlying this form of weak emergence is that weakly emergent, high-level entities may be stable with respect to some lower-level changes (degrees of freedom). This type of idea is further developed by Knox (2016) and Franklin and Knox (2018). So, all of this can be made much more rigorous. (Wilson herself does so as well, using the powers-based subset strategy; for discussion regarding this strategy, see Gillett 2010, Wilson 2011, and Tahko 2020.) But instead of going through the various abstract frameworks, I shall, in Sections 3.2 and 3.3, attempt to illustrate how all this is supposed to work with two case studies.

Before we get to the case studies, let me briefly return to the initial question of this section: the combination of unity and pluralism. What I consider to be the best way of reconciling them has been anticipated in some much earlier work as well. I have in mind the type of 'non-reductive unity' outlined by Harold Kincaid (1990). Kincaid's main concern is a little different – he is mainly interested in showing that molecular biology is not reducible to chemistry. This conclusion is supported by the usual suspects of multiple realisability and broader context-sensitivity of the realisation base. Moreover, Kincaid (1990: 576) conceives of the relevant notion of reducibility in terms of theories linked by bridge laws, which is an approach that we have already questioned, since it misconstrues the relevant sense of ontological reductionism at hand. For instance, he suggests that 'if the goal of reduction is to completely replace higher-level explanations by those at the lower-level, then it is essential that lower-level explanations proceed in entirely lower-level terms' (Kincaid 1990: 577). Interestingly, just as Strevens targets (contrary to what he claims) a very weak sense of irreducibility, Kincaid has here instead picked up, as his target, the strongest possible sense of reducibility, which is incompatible even with semantic anti-reductionism and hence non-eliminative semantic disunity. (He cites Rosenberg as someone who may be committed to this type of strong reductionism, i.e., eliminative semantic unity; cf. Rosenberg 1985, 1994, 2006.) This seems to be a repeating pattern in the literature surrounding reductionism and unity of science: if one targets extreme forms of either reductionism or pluralism – or indeed confuses ontological and epistemic/ pragmatic models of unity – then one can find an easy target. But in any case, it is not this aspect of Kincaid's proposal that I am presently interested in. Instead, I would like to focus on Kincaid's positive story about the unity of science. He pitches this as 'unity without reduction', but of course we must keep in mind, again, that it is eliminative semantic unity and the associated semantic reductionism that he seems to have in mind, whereas reductive ontological unity is sidestepped.

Here is the core of Kincaid's positive proposal: a non-reductive unity of science is based on 'interconnection and dependencies between theories that nonetheless do not allow for one theory to replace the other' (Kincaid 1990: 589). While he formulates this idea in terms of theories, the more detailed outline he provides makes it clear that it is the actual entities described by these theories that stand in the relevant dependence relations. For instance, the very first potential 'interconnection' he describes states that every entity described by one theory may be composed of entities described by the other. This is straightforwardly compatible with the type of compositional reductionism that Gillett has developed. Other types of interlevel dependence that Kincaid mentions include supervenience, heuristic dependence, and confirmational dependence, some of which do not directly concern the underlying ontology, given that, e.g., heuristic dependence has strong epistemic elements. But the key point is that these ontological interlevel dependencies do not entail eliminative semantic unity, and hence all of this seems to be compatible with a type of pluralism about higher-level phenomena.

The upshot of these various attempts to find room for higher-level sciences and hence a type of pluralism while maintaining unity of science (and a form of ontological reductionism) is that much of the work to be done concerns the various dependence relations which together unify the various sciences. These dependence relations are the ontological framework that provides the full picture for the unity of science, but the complexity of this framework is what simultaneously enables the relative autonomy of some higher-level phenomena (with respect to some lower-level degrees of freedom). Let me now put this new outline for a framework for the unity of science, in the spirit of new reductionism, combining reductive ontological unity and non-eliminative semantic disunity, to some use.

3.1 Biochemical Kinds and Special Science Laws

I will consider biochemical kinds such as proteins as the first example.[9] These kinds are at the intersection of biology and chemistry, making them an interesting case of potential cross-cutting and/or reduction/emergence (for discussion, see, e.g., Slater 2009, Tobin 2010b, Goodwin 2011, Bartol 2016, Havstad 2018). Proteins are macromolecules, and if macromolecules are considered to be chemical kinds, we would typically explain their chemical properties in terms of their physical structure, in line with traditional microstructural essentialism. But when considered to be biological kinds, the role of these molecules in physiological processes is important. This role is typically called the (biological) function of the kind. The function is also tied to evolutionary or aetiological considerations, given that biological functions are the result of a causal sequence of evolutionary processes.[10] The key question is of course the relationship between the complex biological aspects, which also involve extrinsic, historical properties, and the microstructural, chemical aspects. For instance, one might suggest that a protein's three-dimensional structure supervenes on its amino acid sequence. This idea can be illustrated with simple

---

[9] This discussion builds on Tahko (2020).

[10] I will omit a discussion of the complicated case of aetiological properties here, but see Tahko (2020) for a case study involving this aspect as well.

textbook models of protein structure, such as the structure of haemoglobin in Figure 3.

But even though there is clearly a compositional dependence relation between the primary structure (amino acid sequence) of a protein and the resulting (tertiary and quaternary) three-dimensional structure, the exact relationship between these aspects is much more complicated. Indeed, many have argued that protein folding is a challenge even for weak forms of ontological unity, and accordingly provides a strong case in favour of disunity or pluralism (e.g., Dupré 2012: ch. 9; Bartol 2016, Havstad 2018). Yet not everyone thinks that this case undermines ontological reductionism (OR) as we defined it at the beginning of Section 3 (e.g., Goodwin 2011, Tahko 2020).

William Goodwin's case in favour of ontological reductionism draws on an analogy with organic chemistry, where we may classify molecules into distinct functional types, not unlike proteins. Organic molecules can be classified depending on their molecular and reactive environments, but their behaviour in these environments is nevertheless considered to result from molecular

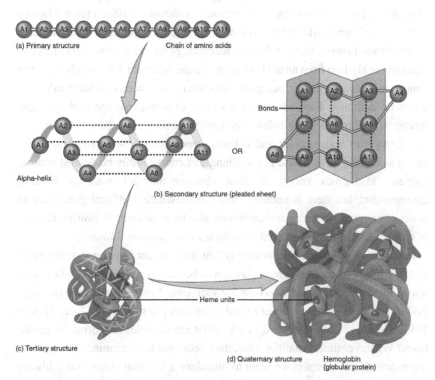

(a) Primary structure — Chain of amino acids

Alpha-helix

(b) Secondary structure (pleated sheet)

Bonds

(c) Tertiary structure

(d) Quaternary structure — Hemoglobin (globular protein)

Heme units

Figure 3 Textbook model of haemoglobin.

structure (which should also be conceived of three-, or in fact four-dimensionally because the structures are dynamic in time). In other words, chemical kinds, for better or worse, are just as 'messy', to use Joyce Havstad's term, as biochemical (or biological) kinds (on 'messy' lab work, see Schickore 2008: 329; see also Rosenberg 1985: 76). As Goodwin (2011: 543) puts it: 'The tertiary structure of a protein, when there is one, represents the most energetically stable conformation available to the protein in the relevant biological circumstances.' So, the process of protein folding, which can produce different tertiary structures from one primary structure, may be understood as being governed by complex interaction with the environment, aiming towards the most energetically stable conformation.

To illustrate that chemical kinds as well are 'hostage' to their environment, it might be helpful to briefly discuss a classic example of a rather problematic pair of chemical kinds: acids and bases (which are discussed in a slightly different context, e.g., by Stanford and Kitcher 2000: 115–119). Acids were historically defined in terms of their macroscopic, phenomenological properties, such as their sour taste and corrosiveness. But chemists have struggled to come up with a proper definition of acids and bases, one reason being precisely contextual variation (on the history of acids, see Chang 2016: 41ff.). The most commonly used account derives from the Brønsted–Lowry theory: acid-base reactions generally involve the transfer of a proton (an H+ ion) from an acid to a base. Hence, we could define acids as proton donors and bases as proton acceptors. However, the obvious problem in giving a general definition for acids is that a variety of substances can undergo these reactions and indeed some substances (called amphoteric substances), such as water, can react both as acids and as bases, depending on their environment. So, acting as an acid or as a base is something that cannot easily be defined without reference to a given context. A given substance may act as an acid in most environments, but there is nothing about the substance itself that guarantees its behaviour as an acid, the environment must also be considered. Following this line of thought, acids and bases could be understood as *functional kinds*.

An interesting feature of this case is that the relevant laws concerning acids and bases, like all special science laws, will be *ceteris paribus* laws – laws that allow for exceptions outside normal conditions. Now, let us grant (as, e.g., Fodor assumes) that laws do not need to be exceptionless to be 'real' (Fodor 1997: 162, fn. 2). We still need to ask: what are the relevant normality conditions? Whatever these conditions are, they better not be determined, say, just by the experimental setup. If we were to introduce a law that states 'All acids are proton donors' based on experiments involving only water as the base, then the ceteris paribus clause would rule out water itself being an acid. In fact, this used to be the case with the definition of acids in the late 1800s, when the Swedish chemist Svante Arrhenius defined acids as compounds that increase the concentration of H+ ions that are present when added to water. Since water was used as part of the normality condition, it is obvious that water itself could not be classified as an acid or base according to this definition. Moreover, any laws to be derived based on this theory would not be able to account for cases where the reacting substances do not contain the relevant hydrogen ions. So, the correct ceteris paribus clause in this case would instead have to be based on the underlying chemical properties of the participating substances, and the Brønsted–Lowry theory enables this, since it is no longer necessary that a substance should be composed of hydrogen (H+) or hydroxide (OH-) ions in order to be classified as an acid or a base.

This brief example serves as our first warning sign: we should recognise that there could be differences in lower-level properties (here, the chemical properties of substances participating in acid-base reactions) that could make a difference in how we classify higher-level kinds (here, the classification into acids or bases). Otherwise, we may end up arbitrarily ruling out relevant lower-level differences based on nothing but the fact that we have not conducted any experiments where these differences would be observable. Therefore, an appeal to special science laws as the motivation to postulate distinct higher-level kinds is not enough on its own. I have previously labelled this as 'the problem of lower-level vengeance' (Tahko 2020).

The account that is emerging here plays into the hands of the ontological reductionist because higher-level kinds turn out to be dependent on lower-level kinds in precisely the way that the ontological reductionist claims. The fact that special sciences generally involve ceteris paribus laws introduces a prima facie challenge for the anti-reductionist, because the relevant normality constraints may also require reference to lower-level kinds, hence undermining the autonomy of the special science laws. Regarding the unity of science, the upshot is clear: We have to consider dependence relations all the way down to make sure that we have the full picture, to determine the complete set of identity-criteria of a given natural kind. These dependence relations are important for ontological models of unity, as they help us determine whether the relationship between higher-level and lower-level phenomena is ontologically reductive or non-reductive.

3.2 The Case of Haemoglobin

As a somewhat more detailed case study, I will consider the haemoglobin molecule.[11] Haemoglobin binds and transports oxygen in blood, releasing it

---

[11] This example is partly inspired by Max Kistler's (2018) recent work. Goodwin (2011: 535) mentions the case in a footnote, and Tahko (2020) goes into further detail.

in cells. Roughly speaking, the biological function of haemoglobin is its ability to bind and release oxygen – more accurately, to carry oxygen from the lungs to the tissues and return carbon dioxide from the tissues to the lungs. The haemoglobin molecule is constituted by four polypeptide chains: two alpha chains (alpha helices) and two beta chains (pleated sheets), which have different sequences of amino acid residues but fold up to form similar three-dimensional structures (as in Figure 3). Only nine of the amino acid residues per globin chain in functionally identical haemoglobin molecules are present in all (functionally identical) haemoglobin molecules. Variations elsewhere in the sequence do not alter the functional profile. So, it appears that we are dealing with a case of multiple realisation, where several different microstructures can produce the same biological function.

In a biological context, we can take two distinct macromolecules (polypeptides) as two instances of haemoglobin, whereby both macromolecules are effectively treated as the same protein, despite microstructural differences. Hence, 'haemoglobin' might serve as a label for a class of proteins defined not only in terms of their microstructural features but also in terms of their functional profile.[12] But we need some more detail about protein structure to assess the example. It is at the level of tertiary protein structure, the three-dimensional combination of alpha and beta chains, where we see the biological function of the protein. There can be several different primary structures that produce the same functionality. This is a common feature of proteins. Significantly different primary structures can produce very similar three-dimensional structures, sufficiently similar in order to produce the same functionality. This is true in the case of the various haemoglobin molecules as well. One aspect of variation in haemoglobin concerns its affinity for oxygen. For instance, since foetuses cannot breathe for themselves, the haemoglobin molecule active in foetal blood has a very high affinity for oxygen (there are three different haemoglobins in effect at different stages of foetal development). After birth, this affinity decreases. While it is important that haemoglobin has a high affinity for oxygen, it is also important that the oxygen gets released when needed. Accordingly, it is not surprising that the haemoglobins between adult mammals of different species may bear more similarities with each other than those between the adults and foetuses of the same species. In fact, haemoglobin can modify its affinity for oxygen depending on its chemical environment, since the adaptability of haemoglobin is tied precisely to its environment (compare this to the case of so-called moonlighting proteins, discussed in Tobin 2010b).

---

[12] Compare this with Bird (2018b), who argues that evolved functional properties may be considered to be macro powers.

One property that influences haemoglobin's affinity to oxygen is the pH value of the environment.

The question that the ontological reductionist faces is whether we can account for the capacities of the haemoglobin molecule in terms of its primary structure. If these capacities are present only at the level of tertiary structure, then this would appear to be a case of multiple realization which is problematic for the ontological reductionist. But why should one think that various primary structures do not all have the same functional capacities? Here we ought to recall Goodwin's point: the case of proteins is analogous to that of organic molecules that may be classified according to distinct functional types. One way to think about these capacities is in terms of dispositional properties that may or may not manifest. The case of acids and bases which we discussed in the previous section is a good example. The initial problem with the simple idea that acids are proton donors and bases are proton acceptors is that a variety of amphoteric substances can act both as acids and bases, and this variation will depend on the environment. Yet it would be odd to claim that the *capacity* of an amphoteric substance to react as either an acid or a base is something over and above its microstructural properties, something that the substance gains only when the relevant environmental circumstances are in place. The thought here is that the capacity to behave in certain ways in different environments can be fully analysed in terms of the intrinsic properties of the substance. After all, both losing a proton (reacting as an acid) and gaining a proton (reacting as a base) are simple chemical reactions that are determined by the relevant molecular structures and we understand these reactions very well. There is no reason to think that we could not accurately capture acid-base interactions in terms of molecular structure, as the capacity to act as an acid or a base is contained, *dispositionally*, in that structure.

We are now in a position to give an analysis of one important aspect of haemoglobin's functional profile in the lines of reductive ontological unity. Amino acids, which have a carboxylic acid group and an amino group (base), are also amphoteric. As we know, proteins are made up of amino acids, and accordingly they can also react as amphoteric substances. So, it turns out that at least some of the interesting capacities of proteins, some of their functional promiscuities, derive precisely from their amphoteric nature, which we have just explained in terms of molecular structure. This looks like a straightforward case of ontological reduction. So, the upshot of this case study is that the biological function of proteins like haemoglobin may, at least partially, be reduced to their microstructure. Indeed, in the case of haemoglobin, its amphoteric nature is in an important role when it comes to oxygen equilibrium; if we want to explain the oxygen affinity of haemoglobin, we will ultimately have to refer to its amphoteric nature.

So, it looks like we can give an analysis of this case that could satisfy the ontological reductionist. This contrasts with, for example, Jordan Bartol (2016: 543), who recommends that we should adopt a 'dual theory' of chemical and biological kinds with respect to macromolecules such as proteins, driven by issues surrounding multiple realisation. This would result in non-reductive ontological disunity and entail a type of ontological pluralism about kinds. Note however that I do agree with Bartol when he points out that there could be *pragmatic* reasons to adopt different classificatory practices, so nothing we have seen above rules out pluralism modelled in the sense of non-eliminative semantic disunity. Continuing this line of thought, Havstad has argued that the primary structure of proteins is not sufficient to account for protein individuation in biological practice. She thinks that we need to consider the aetiological constraints as well and that these constraints are difficult to account for in microstructural terms. (I discuss this issue in more detail in Tahko 2020.) Moreover, even if we knew everything there is to know about evolution and were able to reconstruct this in microstructural terms, we might nevertheless prefer a taxonomy of the relevant biochemical kinds driven by considerations of biological practice. So, does this support disunity/pluralism? Yes, but arguably only of the epistemic/pragmatic type that enables semantic pluralism and can be reconciled with ontological reductionism. Let me now move on to another case study.

3.3 The Case of Electronegativity

This case study comes from the chemistry-physics interface. The question is, can we give an ontologically reductive analysis of chemical properties in terms of physical properties? We should start by making an observation about the distinction between chemical and physical properties quite generally, as this distinction is by no means uncontroversial. For instance, in the nineteenth century, the boiling and melting point were still considered to be physical properties, not necessarily connected with the chemical properties of a substance (Needham 2008: 66–67). Moreover, Paul Needham notes that '[i]t was sometimes taken as an argument against the reduction of chemistry to physics that physics couldn't possibly provide for the chemical characteristics of substances' (ibid., 67). The situation changed once atomic numbers were understood as encompassing the crucial guidelines for chemical properties. Perhaps the simplest case that we can consider here is the reducibility of the chemical properties of a molecule to the molecule's proper parts (and their properties), the elements and subatomic particles that compose the elements (for relevant discussion, see Le Poidevin 2005, Hendry 2010, van Brakel 2010, Needham 2011, Seifert 2017, 2019, and Tahko 2012, 2015b).

There are good reasons to start with the most discussed example in philosophy of chemistry: water. Water has special importance for us as living organisms, but it also has many interesting chemical properties despite its relatively simple structure. We also have a relatively good understanding of what the corresponding physical properties responsible for chemical properties like the boiling and melting point (of water) are. Water's atypically high boiling point is explained by *hydrogen bonding*. When several water molecules are present, they form hydrogen bonds with each other, resulting in a network of hydrogen bonds. Similarly, many of water's other interesting properties, such as its surface tension, derive from the polarity of the water molecule and the resulting bonding behaviour. All this is at the level of chemistry, but we can further analyse hydrogen bonding in terms of *electronegativity*. A hydrogen bond between two molecules forms when two electronegative atoms, such as hydrogen and oxygen, are close enough to attract each other. Electronegativity, according to the standard definition formulated by Linus Pauling, is the chemical property that acts as a measure of the tendency of an atom to attract a bonding pair of electrons. Figure 4 illustrates the electronegativity of oxygen (3,44) and hydrogen (2,20) atoms in an H_2O molecule. Figure 5 illustrates hydrogen bonding between H_2O molecules.

This is where things start to get interesting. Electronegativity, which is generally understood as a *chemical* rather than a physical property, does not have an obvious reductive explanation in terms of more fundamental, physical properties, at least not in the sense that we could give a specific set of physical properties and identify these with electronegativity. There are two issues that we should consider here:

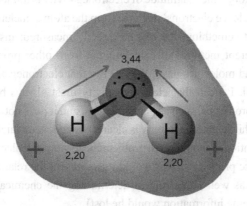

Figure 4 Permanent dipole of water molecule.

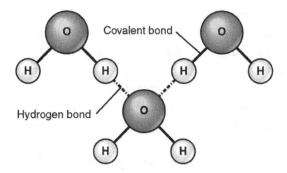

Figure 5 Hydrogen bonds between water molecules.
Source: OpenStax (2013): 'Anatomy and Physiology' (CC-BY 4.0), available at https://openstax.org/books/anatomy-and-physiology/pages/2-2-chemical-bonds.

1. Electronegativity is a *relational* property. It describes the magnitude of the interaction between an atom's valence electrons – electrons on the outer shell – with those of another atom.
2. As we already saw in the case of water, electronegativity is mainly relevant as a property of an atom as *part* of a molecule.

Both of these issues complicate the analysis of electronegativity, although they do not of course block ontological reduction entirely, since the configuration of the valence electrons clearly has a key role. We might say that whatever electronegativity is, it somehow derives from this configuration of the valence electrons. But the point is that we would struggle to replace the notion of electronegativity in chemistry with the configuration of valence electrons. An additional point of interest is that although the magnitude of electronegativity is influenced by atomic number and the valence electrons' distance from the atom's nucleus, electronegativity itself is not something that can be directly measured; instead, there are a variety of different methods of calculating it from other properties that the relevant atoms and molecules have. When I say that electronegativity cannot be directly measured, I mean that electronegativity must always be measured in relation to other properties, such as bond energies, and it does not have a unit. All this means that while electronegativity is clearly dependent on nuclear charge and electron configuration, it would seem problematic to simply replace it in chemical practice with these properties. This would misrepresent the relational aspects of electronegativity as well as its role in determining the chemical properties of molecules (i.e., some information would be lost).

This result speaks clearly in favour of the semantic anti-reducibility of properties such as electronegativity, and hence non-eliminative semantic disunity on the side of epistemic/pragmatic models of unity. More speculatively, on the side of ontological models of unity, this gives rise to the question of whether electronegativity could be an emergent property, introducing causal powers irreducible to fundamental physical properties, which would entail non-reductive ontological disunity. At the very least, the idea that we can see emerging from this discussion is that there plausibly *could* be chemical properties that are, in a sense, *fundamental*, due to being irreducible to more fundamental physical properties. And this is the case even though we know that there are more fundamental physical properties that influence, and perhaps determine, the relevant chemical property. Moreover, even though the presence of these more fundamental physical properties may be necessary for the relevant chemical properties, it might not be sufficient. It is not sufficient for the emergence of chemical properties such as electronegativity that all the relevant lower-level properties are present, since we have just observed that electronegativity is mainly relevant only as a property of atoms when they are a *part* of a molecule. So, as we have seen in other examples, we cannot define such properties without reference to their environment.

One question that we might like to ask, prompted by this case study, is how much do we need to know in order to conclude that some higher-level property is ontologically reducible? None of the examples that I have mentioned throughout this Element have demonstrated anything near a *complete* ontological reduction. But at some point, when we know enough about the dependence relations between a higher-level property and some set of lower-level properties, we seem to be in a position to say that a complete ontological reduction is possible or at least very likely, even if we were not yet (or ever) in the position to state all the relevant bridge laws. However, and in contrast, not just *any* dependence relation is enough. For instance, in the case of electronegativity, what I have said previously is hardly enough for a complete ontological reduction, even though we can point to a clear (compositional) dependence between electronegativity and the relevant valence electrons. In addition, we would need a much more detailed account of the underlying quantum chemistry (for discussion, see, e.g., Llored 2012). Still, the reason for favouring ontological reductionism is clear: so far, we have not encountered clear violations of it. For the friend of ontological unity, this is of course very good news: we can salvage as many higher-level predicates as we like, but at the end of the day, we know it to be very likely that there is a complex network of dependence relations that connects all of science, even if we cannot list all of these relations, and hence cannot eliminate the higher-level predicates – we can still be semantic pluralists.

I will conclude this section by noting that there is another line of argument available that challenges reduction at the chemistry-physics interface. This argument was introduced by Needham (2011). Needham argues that there are no good reasons to accept *microessentialism*, the thesis that chemical substances must be characterised in terms of their microstructure. Instead, Needham proposes that we characterise chemical substances in terms of their macroscopic properties. Needham is mainly reacting against natural kind essentialism based on intrinsic (non-relational) properties familiar from the work of Saul Kripke (1980) and Hilary Putnam (e.g., 1975). Accordingly, I do not think that the argument is quite as damaging against more developed forms of natural kind essentialism (as discussed in Tahko 2015b). The important point here is the need to characterise chemical substances at least partly macroscopically. Philosophers of chemistry have long resisted natural kind essentialism's poster child, namely, the over-simplified identification of water and H_2O. Needham's central message can be captured with this striking quote: 'A scientific characterisation needn't be a microscopic characterisation' (2011: 7; see also Williams 2011 on the intrinsicness requirement; this point has more recently been bolstered in Havstad 2018). But once we realise that the unity of science need not be supported by a commitment to such a strong form of reductionism, it should be clear that the question of microstructural versus macrostructural approach to chemistry does not settle the issue about ontological unity. Indeed, there may still be some hope for a qualified version of microstructural natural kind essentialism even if we do take Needham's arguments onboard – namely, a version acknowledging that the context and environmental interaction of chemical substances need to be taken into account.

3.4 Unity of Science and Ontological Reductionism

The hope for unity of science was originally part of a logical empiricist dream of explicit eliminative semantic unity and implicit reductive ontological unity, the search for a simple relation of explanatory dependence, both at the epistemic and the ontological level. We have known for a long time that this dream was overly ambitious and indeed quite implausible, given the complexity of science and the need for semantic pluralism. But the notion of the unity of science should not remain tied to this long-abandoned endeavour. Given the complexity of science and indeed our own psychological limitations, we need to be open-minded about the various models, methodologies, and theoretical concepts employed in science. In this sense, pluralism – the disunity of science – is obviously alive and well. However, I hope that for those who have read the book this far, it should also be clear that relatively few arguments raised against reductionism and the unity of science, and in favour of pluralism, target the *ontological* dimension of unity. And even when the arguments are about ontology, I have tried to demonstrate that there is much controversy about whether the arguments establish that there is genuine ontological disunity or just epistemic shortcomings or pragmatic issues deriving from scientific practice. None of this is supposed to downplay the importance of such epistemic and pragmatic issues, but we should not be misled about what these issues amount to: they concern the ways in which we *do* science, not the underlying ontology of reality.

We may further clarify the two levels of inquiry that are intertwined here by focusing on two explanatory goals. One of these goals is at the level of 'surface discourse' (i.e., it is primarily pragmatic). I take the notion of 'surface discourse' from Goodwin (2011: 535), but I should note that Goodwin does not appear to use this notion of surface discourse in any pejorative sense. Rather, the idea is that the necessary preservation of certain higher-level predicates at the level of surface discourse suggests a commitment to semantic anti-reductionism, whereby the predicates of the special sciences can still serve an invaluable, even indispensable role (see Gillett 2007: 195, 2016: 16).

Given the important role of the special science predicates, eliminative semantic unity now seems like a non-starter. But the *ontological* reductionist only insists that it is, at least in principle, possible to find lower-level truth-makers for the higher-level, special science talk, where 'in principle' can mean that we will never actually be able to do this, even if science were to be completed. Why is it so important to appreciate this difference? The reason for this is something that John Heil has captured in a very striking way: if we do not properly distinguish ontological matters from the semantic and epistemic import of science, there is a risk that we end up 'reading off' features of the world from features of our language (Heil 2003b: 218).[13] This does not mean that we should not pay attention to scientific practice or map the various competing models that science employs. Even when a number of these models turn out to be incompatible, they may nevertheless all turn out to be valuable in some way (cf. Mitchell 2002,2003). But there is just one world, and insofar as we do not find any genuine cases of strong emergence, then there is some consistent set of worldly dependence relations that ultimately underwrites all these models. Our models may turn out to be abstractions or idealisations that only capture some aspects of those worldly dependencies (cf. Knox 2016) – and this is perfectly fine – but none of this undermines ontological unity.

Let me conclude this section with a conciliatory note. Even if we must be ontological reductionists to salvage the (ontological) unity of science, this does

---

13 See also Ney (2010: 442), where she makes the point that a reductionist can consider higher-level phenomena to be 'abstractions from concrete microphysical situations'.

not take anything away from the practice of higher-level sciences like biology or chemistry. More generally, the status of a scientific discipline does not depend on there being a set of genuine, ontologically irreducible higher-level kinds or special science laws that we can associate with it. However, the ontological reductionist recognizes that the higher-level goings-on depend on the lower-level goings-on in some important respects even when there is higher-level autonomy with respect to certain degrees of freedom (cf. Wilson 2010). This means that even if higher-order scientific language includes some indispensable classificatory practices or natural kind predicates, this does not entail ontological pluralism about natural kinds.

The upshot is a modest type of ontological reductionism that is compatible with a similarly modest type of semantic pluralism. But there is one big issue that remains open, now that I have argued that we are still entitled to pursue ontological unity: where does that ontological unity stem from? I have hinted at a few alternatives, such as Gillett's compositional reductionism, which suggests that we account for the ontological status of higher-level entities in terms of their component entities (but note that as Aizawa and Gillett (2019) argue, we should be pluralist about those compositional explanations as well). But this still leaves the ultimate ontology of those component entities open, and indeed we have many, many options regarding that ontology. My preferred strategy for laying out this ontology is by looking at the notion of a natural kind. This choice is easy to justify, since we saw that the notion of natural kind was central already in the original debate about unity between Fodor and Kim. But now we have the tools to make the underlying ontology much more precise.

In the next section, I will argue in favour of a unified theory of natural kinds – natural kind monism – as this enables us to preserve a core motivation for the old search for the unity of science as well.

4 Unity of Science and Natural Kinds

In Section 2 we saw that the influential debate between Fodor and Kim led to the normalisation of pluralism: the phenomenon of multiple realisability and failure of eliminative semantic unity are widely taken to show that there are plausibly at least some genuine higher-level natural kinds. In Section 3 we saw that despite the dominance of semantic pluralism, there is still room for ontological unity. What remains to be done is to give an account of the ontological basis of that unity. I propose that this can be best done by adopting *natural kind monism*:

Natural kind monism: The view that there is a single notion of 'natural kind' and anything falling under that notion can be defined in terms of the same general set of identity-criteria.

The first obvious question about this view is whether we should link it to the ontological models of unity and hence the structure of reality or whether it links to epistemic/pragmatic models of unity and the structure of scientific theories. I intend the first reading: natural kind monism should be understood as an ontological position about what the notion of 'natural kind' refers to within the structure of reality. Hence, the view suggests that we need to postulate just one fundamental ontological category to account for natural kinds. This still allows us to accommodate semantic pluralism about higher-level kinds. When we ontologically reduce a higher-level kind to a lower-level kind, we are dealing with one set of properties, the combination of which realises the higher-level kind – that is, in the line of ontological reductionism, the truthmakers for higher-level-kind talk are to be found from lower-level kinds. To clarify how I understand this view, let me cite a passage from a recent paper by Jordan Bartol, where he proposes a distinction between two different senses of 'monism'. Bartol is here interested in whether microstructuralism, the idea that kinds are defined in terms of their (microstructural) parts, could provide a workable sense of natural kind monism and indeed of the ontological unity of science:

> First, microstructuralism offers monism in the fashion after which molecules are naturally individuated. Call this 'category monism' (CM), since it is monism about the ontological category, 'kind'. Microstructuralism holds that all chemical kinds are what they are in virtue of microphysical facts. Contrast this with functional kinds, historical kinds, or relational kinds. If these other brands of natural kind were admitted to our ontology then we would have a plurality of kind categories. Second, microstructuralism offers the promise of a single taxonomy. Call this 'taxonomic monism'. Every kind in the microstructuralist taxonomy is unique. (Bartol 2016: 537)

It is clearly 'category monism' that we are focusing on, since 'taxonomic monism' could be understood merely as a pragmatic or semantic thesis. So, natural kind monism is a promising way to systematise my preferred understanding of ontological unity, namely, reductive ontological unity.

So, what could a monist account of natural kinds look like? One popular approach is natural kind essentialism (e.g., Ellis 2001, Tahko 2015b), but natural kind monism should not be considered to entail an essentialist account. Bartol himself focuses on microstructuralism, but monism can take many other forms. Note however that I would not agree with Bartol that microstructuralism necessarily promises a single taxonomy. One can be a monist about the category of natural kind but still believe that we need to (or at least can) be *taxonomic* pluralists (cf. Tahko 2020). This is the lesson that I have been trying to make clear throughout this Element: reductive ontological unity is compatible with non-eliminative semantic disunity. For instance, proteins may be classified in terms of their function, aetiology, structure, or some combination of these, but this does not rule out a monist approach to their ontological status. Bartol also associates monism closely with reductionism and this is fine insofar as we are talking about ontological reductionism and keep in mind that the view is compatible with theoretical or semantic anti-reductionism and hence the indispensability of higher-level predicates (as noted by Gillett 2007 and 2010: 188).

In contemporary literature on natural kinds, various forms of pluralism dominate the discussion (e.g., Dupré 1995, Cartwright 1999, Mitchell 2003, Chang 2016, Magnus 2012, Khalidi 2013, Waters 2016). A detailed analysis of these various versions of pluralism is not possible here, but it should be noted that some of them are *ontologically* pluralist, in contrast to the semantic pluralism that we have been discussing earlier. While semantic pluralism is compatible with natural kind monism, ontological pluralism (e.g., Bartol's [2016: 543] 'dual theory' of chemical and biological kinds) is clearly not. Instead of attempting to do justice to the various version of pluralism about kinds, it may be helpful to take a step back and discuss the more general issue of *realism* first. The reason for this is that I believe that there are aspects of the contemporary debate about natural kinds that have skewed the debate towards ontological pluralism about natural kinds.

To understand the debate between pluralism and monism, we need to first clarify that there is a shared sense of realism. If natural kind monism is supposed to lead to the ontological unity of science, then it must be committed to realism. Note that at least some of the pluralist approaches mentioned earlier would not share this commitment to realism (perhaps, e.g., Magnus 2012, Chang 2016, Waters 2016), since these are sometimes combined with explicit pragmatism or anti-realism. But then the debate is really about realism/anti-realism rather than about unity/pluralism as such, so we must set it aside here. Recall that ontological unity concerns ontological rather than epistemic or theory reduction. The reason why this is worth pointing out is that while any account of natural kind monism is plausibly realist, taxonomic monism need not necessarily be realist. So, the first question is: what does realism about natural kinds amount to, and is there a shared understanding of this realism between monists and pluralists? We can clearly see that if the answer to the second part of this question is negative, then we need to first settle the question of realism before we even get to the question of kindhood.

As a working hypothesis, natural kind realism may be understood as the relatively weak thesis that there are entities – the natural kinds – which reflect *natural* divisions in mind-independent reality. When laid out in this rather ambiguous fashion, natural kind realism is a widely shared view and perhaps only opposed to the conventionalist view that the classifications we postulate *never* reflect any natural divisions in mind-independent reality. However, the interpretation of 'natural' is left open here, with the result that natural kind realism encompasses several mutually inconsistent views.[14] We may, however, further specify the view on a very general level. A potentially helpful clarification – even if it has been overused – is that natural kind *terms* 'carve reality at its joints' (cf. Khalidi 1993). Accordingly, natural kind terms may be thought to be the set of *concepts* that we use to refer to the mind-independent 'joints' of reality. These concepts may carve reality to a varying degree of accuracy – and sometimes we may be mistaken about whether a concept successfully carves – but in order to avoid ambiguity, we can start with the idea that at least the concepts that carve nature at its joints 'perfectly' constitute natural kind terms.

One initial source of concern is that the notions of 'naturalness' and 'perfectly natural' have not been specified. Another source of concern, brought to light already by John Stuart Mill (1843/1882) is the distinction between natural *kinds* and mere groups of objects characterised by a shared natural property – for example, chemical substances constitute a fairly plausible example of kindhood, whereas being acidic, or being green and round, are more plausibly merely shared *natural properties*.[15] Thus we have to distinguish between the idea of grouping some set of entities together on the basis of one or more shared properties, and the idea of grouping them together because they are of the same kind. It *may* be a minimal, necessary criterion for shared kind membership that two entities share at least some properties (taking note that many would deny even this, especially in cases such as biological species which could be understood purely relationally/historically; cf. Okasha 2002), but it is most certainly not sufficient, because otherwise we would end up reifying all sorts of things as natural kinds (e.g., all green and round things).

A more striking example: electrons and muons are both elementary particles and they have the same charge (unit negative charge) and half-integer spin. But muons have a much greater mass. So, while electrons and muons are indeed

---

[14] The interpretation of 'mind-independence' is not uncontroversial either but should be understood weakly: divisions in mind-independent reality do not depend on our conceptual schemes, or on the presence of human observers. Moreover, the way that conventionalism is laid out above is, of course, simplified and not many people would subscribe to it. For further discussion about extreme conventionalism, see Tahko (2012). See Sidelle (2009) for a defence of moderate conventionalism. I will discuss the mind-independence requirement in more detail in Section 4.2.

[15] See Bird and Tobin (2018), Bird (2018a), and also Hawley and Bird (2011), where an initial contrast between these questions is provided and one account of the 'kindhood question' is presented. Regarding the case of acids, see Stanford and Kitcher (2000: 115ff). The thought here is that there is no shared microphysical basis for different kinds of acids, only a functional role, as discussed in more detail in Section 3.1.

both members of the same higher-level kind 'lepton', and since they both have a half-integer spin, it is clear that they are of two distinct kinds. When we are giving necessary and sufficient conditions for kind membership, we should first consider the *narrowest* kind. If we instead focused on some very broad kind, like all things that have mass and extension, we would not get the desired result. We can certainly make predictions about the behaviour of objects that have mass and extension – for example, on the basis of the influence of gravity on these massive bodies, but insofar as there is a more precise distinction to be made, then we ought to make it. Once we do, we can see that shared properties are not sufficient for shared kind membership. That is why we need an additional constraint, something that explains why the properties are shared in the first place.

We can capture these ideas with the following two constraints, the specification of which will be my main task in the following section:

The naturalness constraint (NC): Members of a natural kind must share at least some natural properties.

The kindhood constraint (KC): The reason why members of a natural kind share a set of natural properties is because they are of the same kind.

Both KC and NC require further analysis before they can be deemed informative. The important work is done by the notions of *naturalness* and *kind*. There have been many attempts to determine what makes a natural kind genuine or real, often in terms of some further unifying factor, such as causal mechanisms or laws of nature. These could be understood as ways of addressing KC – as providing reasons for why certain properties are unified in a sufficiently strong fashion to constitute a natural kind. However, there are subtle issues surrounding this question. My own preferred understanding of KC considers that the kind unifies its properties. Specifically, what explains that members of a natural kind generally share many of their properties is that the kind unifies these properties. It may do so in terms of causal mechanisms or by other means, but the reality of a natural kind is not derived from its *unification principle*. Rather, it is something about the nature of the kind itself that explains why and how it unifies certain properties.

Following this line of thought, KC itself suggests that the kindhood question cannot be settled merely by listing a set of shared natural properties among the members of a natural kind (even if this may be how we arrive, epistemically, to conclude that they must be of the same kind). To repeat, it is a shared kind membership that explains the shared properties (if any), not the other way around. This leaves it open that there *could* be two members of a natural kind that do not have any properties in common, so strictly speaking, NC is not required for the existence of natural kinds, but it is a plausible constraint, so I will assume that it is true. I should note that KC does immediately rule out some reductive approaches to natural kinds – for instance, views according to which there is nothing more to being a natural kind than a shared set of natural properties (see Tobin 2013 and Bird 2018a for discussion).

4.1 The Naturalness and Kindhood Constraints

The starting point of the naturalness constraint (NC) is that what is 'natural' about natural kinds is their correspondence with the mind-independent joints of reality: natural kind terms aim to reflect the structure of reality. There are several ways to cash out the distinction between natural and non-natural properties in this sense (e.g., Lewis 1983). For some, resorting to naturalness is so obvious that arguments in favour of it do not amount to much more than an appeal to 'knee-jerk realism' (Sider 2011: 18).[16] However, one might think that relying on a David Lewis-style account of naturalness will not get us very far if we hope to define natural kinds quite generally, because the focus of the discussion is not so much on kinds, but rather on a *nominalist* account of properties – a nominalist account denies the existence of universals. Yet naturalness may nevertheless be the best candidate we have for clarifying the idea that not all properties are 'on a par', as Cian Dorr and John Hawthorne (2013) put it in their lengthy discussion of naturalness.

Rather than discussing the history of naturalness, we should instead focus on the more pressing question of how we can put natural properties to use when it comes to identifying natural kinds. This is an important issue for any account that resorts to natural properties. Here, the most popular option seems to be *projectibility*. As Tobin puts it: 'The naturalness of kinds makes them projectible; the observation of some instance of a natural kind Ka, licenses the inference that any future instance of K will be similar to Ka and gives credence to the belief that any future K will be similar to Ka' (Tobin 2013: 165).

We first encountered projectibility in Section 2.3, when discussing the case of jade. We can clearly see the important role that it plays in discussions of naturalness and natural kinds. However, the key problem with projectibility is that the inductive inference described by Tobin in this quote can be upheld simply by virtue of shared natural properties between observed cases (cf. also Kornblith 1993). In other words, as we have already seen, projectibility by itself is not enough because it can give us false positives – cases where inductive

---

[16] Sider (2011) presents an influential, Lewis-inspired account of natural properties but does not discuss the kindhood question in any detail.

inferences are upheld merely because of (perhaps accidentally) shared natural properties.[17] Tobin is careful to note that projectibility is only associated with giving some credence to the belief that Ks observed in the future will be similar to the previously observed instance. That's because members of kinds need not be similar in all respects. But we should similarly be wary of cases where members of two distinct kinds share some properties. To control for this, we need the kindhood constraint.

The kindhood constraint (KC) will cause immediate controversy. One reason for this is the variation in the ontological weight that philosophers are willing to admit for natural kinds. But since the kindhood question is here separated from the naturalness question, perhaps we can make some progress. It is worth noting that we should also distinguish the kindhood question from the universalism/ nominalism question.[18] This is despite the fact that I will operate in the realm of realism about natural kinds. We might compare this to the analysis of natural kind realism by Katherine Hawley and Alexander Bird (2011: 206), who simply assume the existence of universals and hence dismiss nominalism altogether, limiting the discussion to those views that accept universals. I will focus on these views as well, but since we have distinguished the kindhood question and the naturalness constraint (NC), it may be possible to account for the relevant type of unity doing the work in KC in some other way than by postulating a universal. So, I think that one could be a realist with regard to KC but a nominalist with regard to NC – that is, to deny that either natural properties or natural kinds themselves should be understood as universals. The idea of unity of science should not be settled on this basis, but this issue will ultimately be faced by any account of ontological unity based on natural kind monism.

Even if we assume the existence of universals, a number of different options remain. For instance, kinds themselves might be considered to be particulars rather than universals, they might be considered as simple or complex universals (the latter option is preferred by Hawley and Bird), or they might be considered as *sui generis* entities, forming one of the fundamental ontological categories. Each of these approaches would give a somewhat different picture of what is responsible for shared kind membership. I will not discuss the details of these options. Instead, I shall focus on my preferred account. I favour the idea that natural kinds form one of the fundamental ontological categories – this would give them the highest ontological weight possible, and indeed I believe that this gives us a very neat sense of unity of science as well. But what does this mean?

---

[17] For a helpful discussion of false positives with regard to the case of natural kinds, see Kendig and Gray (forthcoming: sec. 3).

[18] See Bird and Tobin (2018) for a lengthier analysis of the different options.

Brian Ellis (e.g., 2001) and E. J. Lowe (e.g., 2006, 2015) have provided the best-known accounts of kinds as a fundamental ontological category. Here we will collectively group these views under the label of *natural kind fundamentalism*. This use of 'fundamental' should not be confused with fundamental physics. Lowe specifies the idea as follows: 'What does it mean to describe a certain ontological category as being "fundamental"? Just this, I suggest: that the existence and identity conditions of entities belonging to that category cannot be exhaustively specified in terms of ontological dependency relations between those entities and entities belonging to other categories' (Lowe 2006: 8).

An example may help, so let me borrow Lowe's own example: if particulars consist of coinstantiated universals or perhaps from bundles of tropes, then the category of 'particular' cannot be fundamental in the relevant sense. That's because in that case, the existence and identity conditions of particulars would be entirely specified in terms of the universals or tropes that constitute it (cf. Keinänen and Tahko 2019). Lowe himself would include the category of 'substance' as one of the fundamental categories. It may be helpful to include an illustration of his four-category ontology (Figure 6), which, as the name suggests, postulates four fundamental categories.

We need not dwell on the details of Lowe's ontology, but we may note that the top-left corner of Figure 6 (substantial kind universals) that is of particular interest to us. Relatedly, for Ellis, kinds feature in three of his six fundamental ontological categories, as substantive kinds (generic substantive universals), dynamic kinds (processes, events), and property kinds (Ellis 2001: 97–98, 134). For Lowe, there are two types of universal – substantial and non-substantial – where the first corresponds with substantial kinds (i.e., natural kinds; 2006: 21, 33). So, in both ontologies, kinds feature among the fundamental ontological categories, although for Ellis they are distributed across several categories and

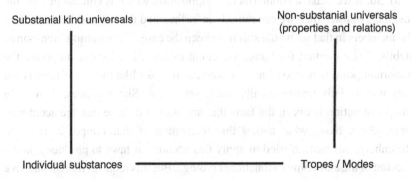

Substantial kind universals ━━━━━━━━━━━━━ Non-substanial universals (properties and relations)

Individual substances ━━━━━━━━━━━━━ Tropes / Modes

Figure 6 The 'ontological square', modelled after Lowe 2006: 22 (author's own work)

hence perhaps better regarded as a subspecies of universals. In any case, I consider both Lowe and Ellis as proponents of natural kind fundamentalism.

Even if kinds were not fundamental entities in this sense, many would nevertheless consider them as *entities* of some sort. (Bird [2018a] calls this view 'strong realism'.) In particular, one might regard natural kinds as emergent entities. The homeostatic property cluster (HPC) theory familiar from the work of Richard Boyd (e.g., 1991, 1999) could be seen as advocating this type of view. In this case, the resemblance among the members of the kind would seem to be derived from the role that kinds play in projectible, inductive generalisations. In other words, the resemblance between the members of a kind is manifested via a shared causal structure; each kind is a property cluster kept in homeostasis by a causal mechanism. But while this type of account has been very popular, it is also clear, and well-known, that the account is unlikely to apply to *all* (natural) kinds (see, e.g., Khalidi 2013: sec. 2.6, Magnus 2012, Slater 2015). For instance, it is difficult to see what kind of causal mechanism could be postulated to be responsible for the resemblance amongst chemical elements: a shared atomic number and nuclear charge are not *mechanisms* in HPC's sense, but rather just individual (possibly essential) properties. The situation is different in the case of biological species, of course, where the HPC theory has proved to be quite popular (but see Khalidi 2013 for some critical remarks about this). In any case, it does seem that the HPC theory could at best offer only a partial account of kindhood – although various novel, more extensive versions of the framework have been proposed (such as Slater 2015; Bird and Hawley 2011 also put the HPC theory to broader use).

Take Matthew Slater's version, where resemblance is derived from Marc Lange's (e.g., 2009) idea of the stability of sets of laws under counterfactual suppositions (Slater 2015: 398ff.). As Slater puts it, the idea is that 'certain sets of truths are maximally invariant under counterfactual perturbations' (ibid., 398). So, if we make a counterfactual supposition which is consistent with the members of a set, then we can reliably infer that, had the supposition been true, the members in that set would still have been the case. The resulting 'non-nomic stability' is important for Lange's account of laws, but for our purposes, the important point is that, on Lange's account, it looks like the set of laws is the only non-trivially non-nomically stable set. So, as Slater notices, there is 'a sharp distinction between the facts that are laws and those that are accidents' (ibid., 399). Now, what makes this interesting is that Lange, in reply to Rosenberg, has himself tried to apply this account of laws to produce a novel account of the autonomy of functional biology. But this requires more restrictive stability than we get with the set of all laws. All this will be easier to understand with an example, so let me cite a passage from Lange, which Slater also cites:

[I]t is of medical interest to know whether a given heart attack might have
been less serious had epinephrine been administered sooner, or had the
patient long been engaged in a vigorous exercise regimen, or had she been
wearing a red shirt, or had the Moon been waxing. But it is not of medical
interest to know whether the heart attack might have been less serious had
human beings evolved under some different selection pressure. A physician
might blame a patient's untimely death on her smoking, but not on human
evolutionary history. (Lange 2004: 107)

So, Lange's suggestion is that we can restrict the relevant range of counterfactuals in terms of the *interests* of the special sciences – in this case, the medical
sciences. Slater reports sympathy towards this basic idea, but he rightly points
out that 'there are some pressing concerns about the details of how interests
apportion modal space' (Slater 2015: 400). This is also the concern that I have,
because the upshot of Slater's account of kinds seems to become, as he puts it,
'domain-relative' (ibid., 403), due to the fact that the definition of stability (on
Lange's account) is dependent on the interests of the relevant special sciences.
Now, Slater prefers to formulate the account in terms of 'relevance' instead of
'interests' (ibid., 401), but the domain-relativeness is nevertheless at the very
core of his account. It should be noted that Slater acknowledges all this – for
example, 'if some kinds are domain-relative, the question of what kinds there
are *tout court* is not generally tractable' (ibid., 404). But it is clear that this will
not satisfy those sympathetic to natural kind monism, because it is precisely
a singular – and realist – sense of natural kinds, *tout court*, that the monist
requires.

Whether Slater's account satisfies natural kind realism is a more controversial
issue, but he does explicitly state, contra Lowe (2006), that on his account,
natural kinds are not an ontological category (Slater 2015: 406). On Kendig and
Grey's (forthcoming) reading, Slater aims at a metaphysically neutral account
of natural kinds – something that they regard as incomplete without explicit
metaphysical presuppositions. Moreover, Slater's account may violate what
I regard to be a minimal criterion for natural kind realism – namely, that natural
kinds reflect natural divisions in *mind-independent* reality. (I will return to the
mind-independence criterion shortly.) Moreover, any account that offers only
a partial solution to the kindhood question will be unable to account for the
ontological unity of science.

One competing account that attempts to provide a more unified approach to
natural kinds and is hence worth considering in this context is the one developed
over many years by Muhammad Ali Khalidi (e.g., 2013, 2018). Khalidi suggests that 'epistemic practices [in science] aim to uncover the divisions that
exist in nature' (Khalidi 2013: 63). Or as he puts it some years later, 'the

ontological ground of projectibility is causation' (Khalidi 2018: 1389). Khalidi's 'simple causal theory' of natural kinds has many benefits over competing views and I find it to be one of the most clearly presented and detailed accounts of natural kinds overall, but it is his take on the kindhood question which I will take issue with. The reason for this is that Khalidi, like many others, ultimately takes projectibility to be the most important guide to kindhood. As we have already seen, this puts a lot of weight on the epistemic practices in science and it is this feature that I believe to have skewed our conception of the unity of science towards the epistemic/pragmatic rather than the ontological. So, there is no question that Khalidi's account is supposed to be realist, but ultimately, any realist account will have to address the ontological dimension as well, and the relationship between properties and kinds is one of the most pressing issues here (see Khalidi 2013, sec. 6.2).[19] Here is how Khalidi first ties these together:

A natural kind K is associated with a set of projectible properties $\{P_1, \ldots, P_n\}$.

As Khalidi notes, this is surely not enough, as we also need an account of what unifies the properties associated with a kind. The problem of finding a reliable link between projectibility or inductive generalisations and natural kinds is of course not new; here we return to the debate between Fodor and Kim discussed in Section 2. Let me cite a passage from Ruth Millikan (whom Khalidi also cites) to illustrate:

[A] natural kind corresponds not just to a projectable predicate, but must figure as the subject of many empirical generalizations. No science consists of a single generalization, nor of a heap of generalizations about different kinds of things. A science begins only when, at minimum, a number of generaliza-tions can be made over instances of a single kind, for example, over instances of silver, or instances of humans, or instances of massive bodies, or instances of, say, moments in the American economy. (Millikan 1999: 48)

Khalidi builds on this idea from Millikan by introducing the requirement that there needs to be 'no shortage' of projectible predicates associated with the kind K. So, the inductive inferences that natural kinds support need to have a certain robust-ness. What explains this robustness? Khalidi's preferred answer is that there is a causal link between the properties. Thus it is this causal 'network of properties' that distinguishes natural kinds from non-natural kinds – Khalidi coins this idea by stating that kinds are the 'nodes in causal networks'. In a little more detail, the

---

[19] Interestingly, Kendig and Grey (forthcoming) argue that even epistemology-only accounts of natural kinds, such as Slater's (2015), will require commitments regarding the metaphysical status of kinds. The same may apply more broadly to epistemology-only accounts of unity.

thought is that there are some 'primary' or 'core' causal properties which are co-instantiated with 'secondary' or 'derivative' causal properties, because the primary properties directly cause the secondary properties. An example of this distinction that Khalidi gives is the case of gold atoms, where 'the primary causal properties of gold include atomic number 79 as well as a disjunction of mass numbers, which give rise in turn to a cluster of other causal properties (e.g. ionization energies, atomic radius, etc.)' (the latter being secondary properties) (Khalidi 2018: 1384). This is a sensible approach in that it allows for certain variation in the secondary properties (for instance, some secondary properties could be masked), while the causally prior primary properties would still be present. Hence, we should identify the kind with these primary properties, even if the kind is also associated with the secondary properties. The key question is of course how the primary properties become co-instantiated. Here, Khalidi is somewhat more tentative, but ultimately his account seems to be based on the idea that the causal networks track laws of nature: 'it seems that all we can say is that some combinations of properties in the universe are allowable and others are not, and that is ultimately a matter of natural law' (Khalidi 2013: 205).

The upshot of Khalidi's account, it seems to me, is that we must reduce the kindhood question into a question about the laws of nature. But this, of course, requires an account of laws, which Khalidi is surprisingly silent about. Moreover, given the major role that causation plays in the account, there is a serious challenge that emerges from the old Russellian line, whereby causation is thought to be absent from the fundamental, microphysical level. This is, of course, a controversial matter, one that we cannot discuss in detail here (but see Khalidi 2013: sec. 6.3 for discussion). In any case, if microphysical causation is denied, this would seem to require one of two things: either we must give up microphysical natural kinds altogether or we must posit two classes of natural kinds and account for the fundamental physical kinds by some other means (an option that Khalidi considers, with reference to Ladyman, Ross et al. 2007). But either option will be alarming for the friend of ontological unity framed in terms of natural kind monism, since they would both entail either giving up natural kind monism or leaving a part of science – and arguably the most fundamental part – out of the picture.

4.2 Mind-Independence, Cross-cutting Kinds, and the Hierarchy Thesis

Before I present my own preferred account of natural kinds in more detail, I would like to briefly discuss some pressing challenges for realism and monism about natural kinds. I shall start with the role of the mind-independence criterion for the naturalness of kinds. I have noted in passing that a common way to conceive of what is 'natural' about natural kinds within natural kind realism is their correspondence with the mind-independent joints of reality (and indeed I have defended this type of view before in Tahko 2012 and 2015b). But some might find this criterion unsatisfactory, primarily because there are supposed natural kinds that are at least partially mind-*dependent*, as we shall shortly see. Indeed, this is a potential threat for the idea that a unified account of natural kinds could truly support the ontological unity of science. For if there are higher-level sciences that are ruled out from this unified account of kinds by definition, then we are clearly not providing a complete unified picture. Now, proponents of natural kind fundamentalism might argue that such supposed higher-level kinds should be ruled out because they are subjective and/or reducible to lower-level kinds. But showing this is by no means straightforward. So, let us consider the issue in some more detail.

Two interesting cases of seemingly mind-*dependent* kinds are non-naturally occurring transuranic elements such as the element with atomic number 99 (*Einsteinium*), and psychiatric kinds such as mental disorders – the latter being mind-dependent by definition.[20] However, I think that we need not worry about such cases that seemingly violate the mind-independence criterion; indeed, the thought that natural kinds need to exist mind-independently has sometimes been misunderstood or misrepresented.

Khalidi (2016: 228) has helpfully listed the following four different ways to understand the mind-independence criterion:

1. Mind-dependence of the kind vs. its instances
2. Causal versus constitutive mind-dependence
3. Contingent versus necessary mind-dependence
4. Mind-dependence versus theory-dependence

It is easy to see that kinds like Einsteinium will satisfy (1), given that the kind itself (say, understood as a substantial kind universal) may be considered mind-independent even if its instances (i.e., the specific Einsteinium atoms that humans have synthesised) are, in a sense, mind-dependent. However, the case of the various psychological and social kinds is more controversial, as one might think that without any minds at all, such kinds could not exist. The details will depend on whether one believes that kinds can exist without any instances – something that I would find questionable, being sympathetic to the Aristotelian

---

[20] I will set aside the complicated case of biological species here, but see Khalidi (2016) for examples involving biological kinds. In the interest of transparency, I should note that I am somewhat sceptical of including biological kinds among natural kinds in the first place, but this is not the place to explore this complex issue.

view that all universals must be instantiated in the actual world (for discussion, see also Hommen, forthcoming).

Regarding (2), the second of Khalidi's suggested ways to understand mind-independence, the thought is that human minds may be causally involved in the creation of certain natural kinds, but these kinds may nevertheless be constitutively mind-independent. Clearly, kinds such as Einsteinium would count as mind-independent on this criterion as well, whereas mental disorders might not. (3) brings in explicitly the idea I mentioned in connection to (1), namely, that certain kinds, such as mental disorders, could not have come about without human minds, whereas we can perhaps allow for the possibility that kinds like Einsteinium could have been created spontaneously in the natural world. (This may be controversial, given the focused efforts required in creating such kinds.) Finally, the distinction suggested in (4) attempts to make the case for mind-independence in terms of ruling out social kinds such as money because their existence is dependent merely on there being a relevant theory (i.e., economics), but Khalidi questions this, noting that it is certainly not the case that all social kinds are entirely theory-dependent.

Indeed, Khalidi himself thinks that (1)–(4) are all problematic. Instead, he proposes a fifth formulation, whereby the distinction between mind-independent and mind-dependent kinds can be specified in terms of whether the instances of the relevant kinds require human mental activity to sustain them as members of those kinds. If such a need exists, the kind is mind-dependent. Even though this is Khalidi's preferred way to understand the mind-independence criterion and it does seem to provide a way to distinguish social and psychological kinds from cases such as artificial elements, his main claim is that this distinction does not provide a good ontological account for distinguishing real or genuine kinds from *ersatz* kinds. More generally, he concludes that mind-independence of the provided varieties should be considered irrelevant to realism about kinds. I beg to differ.

In my view, the point of the mind-independence criterion for genuine natural kinds amounts to this:

Mind-independence (MI): There is a set of *objectively unified properties* that is being tracked by the relevant kind term.

So, what makes a set of properties 'objectively unified'? This concerns the relevant unification principle, the manner in which the properties are unified. But it is, in fact, easier to say what makes a set of properties *not* objectively unified: if the set of properties that a kind term tracks is based on some human interests, then that set of properties is not objective and the kind in question is mind-dependent.[21] It may help to give an example. Consider the phenomenon of radioactive decay. When we say that something is radioactive and classify various isotopes in terms of this radioactivity, we have in mind relatively unstable nuclei. Nuclear stability comes in degrees, but we have never observed proton decay and we classify nuclides as stable when they do not spontaneously undergo radioactive decay. Yet this may all be because we have not had sufficient time to observe such decay – and likely never will. In a famous article, Freeman Dyson suggests that over a time scale of 10^{1500} years, ordinary matter is radioactive – ultimately, all matter will decay into iron (Dyson 1979: 452). But we are generally not interested in such incomprehensibly long time scales, and for good reason. Nevertheless, if Dyson is right, the objective property of 'radioactivity' is much more promiscuous than we ordinarily think – that is, when we have our interest-relative glasses on. Properly speaking, it is not the property of radioactivity that is interest-relative here, but rather the property of 'nuclear stability', as we ordinarily use it (not to be confused with Slater's use of 'stability' discussed earlier). But it is clear that we would lose something if we stopped using the notion of nuclear stability, because for virtually all of our scientific interests, it makes absolutely no difference to us if a nuclide that we classify as 'stable' decays, say, in some 10^{100} years.

The previous example hopefully serves to highlight that just because a kind or a property is mind-dependent does not mean that such a kind could not be useful or that it could not track some causal structures. Rather, such a kind or the associated set of properties satisfies some looser sense of cohesion or unity, say, given a particular epistemic goal (although this is not to say that the goals are *purely* epistemic). If we take the case of synthetic elements like Einsteinium, the proper mind-independence criterion, is easily fulfilled: there is an objectively unified set of properties (e.g., nuclear charge), which is associated with Einsteinium. The fact that the existence of any given instance of Einsteinium is dependent on human action is irrelevant. Moreover, the mere fact that social kinds and psychiatric kinds are mind-dependent is not the problem; rather, problems arise if the set of properties reified as a kind is picked up on the basis of our epistemic interests rather than some objective unification principle, whether or not the properties themselves are mind-independent.

---

[21] I acknowledge that there are complications regarding the notion of 'objectivity', as an anonymous reviewer points out. I use this notion mainly because of its somewhat intuitive association with mind-independence, but I should note that there are senses of objectivity which allow, say, for intersubjectivity. For a comprehensive survey of 'objectivity', see Daston and Galison (2007).

For a natural kind realist, the main issue is that the kindhood constraint (KC) is satisfied and that it is satisfied because the kind is genuine (i.e., satisfies the objectivity requirement specified by MI). There are a number of ways that this genuineness can be captured, but according to my preferred account, genuine natural kinds are substantial kind universals. Such substantial kinds are associated with a given set of properties because the kind *unifies* these properties. Hommen (forthcoming: [3]) puts this nicely: '[K]inds represent unified ways of being – both in an individual and in a collective sense: they account for the modal and temporal stability of character both within single particular objects and across what we then call different members of the same kind.' This may be contrasted with the view that Mill may have held – namely, that at least fundamental natural kinds unify their properties as a matter of brute fact. In other words, there are no causal connections between, say, unit negative charge, half-integer spin, and the rest mass of electrons, even though all three of these are presumably definitive of the natural kind *electron*.[22] This bruteness approach may also be found, for example, in Chakravartty (2007: 171), who suggests that fundamental natural kinds, such as electrons, have their core properties (mass, charge, spin) as a matter of brute fact. Yet this is not the account that Hommen is advocating. It *may* be the case that we cannot point to any causal mechanism or other clear explanation for why certain properties, especially those of fundamental natural kinds, cluster together. But instead of postulating this as brute fact, a natural kind fundamentalist can see this as a justification for postulating natural kinds as a fundamental ontological category. This is the account that Hommen seems to favour, and one that we can also see in the work of Ellis and Lowe. I shall outline this account in the next section, but first I would like to consider one more challenge for natural kind monism in general: the possibility of *cross-cutting kinds*.

The starting point of natural kind monism is that we can account for all natural kinds in terms of the same general identity-criteria. But an important addition to this, suggested by Ellis (2001) is the thought that an entity can only belong to one (fundamental) natural kind. Probably no one is committed to this idea in its strongest possible sense – which Khalidi (2013: 69) calls the *mutual exclusivity thesis*, whereby there really is only a single natural kind that an entity can fall under. This would entail, for instance, that an entity cannot belong both

---

[22] Of course, it could turn out that we will one day discover such causal connections. Moreover, even if there are no causal connections to be found, there could still be something about the nature or essence of these properties that explains why they cluster together (thanks to Sam Kimpton-Nye for highlighting this option). The account that I favour relies instead on the natural kind universal *electron*, which explains the clustering (see also Keinänen and Tahko 2019).

to the kind *proton* and *fermion*, even though all protons are fermions. But it is, of course, very easy to weaken the thesis somewhat, so that an entity may indeed belong to several genuine kinds at once, as long as these are a part of a nested hierarchy. Since all protons fall under the more general kind fermion, it is easy to see that there is nothing here that threatens natural kind monism – indeed, this kind of hierarchy reminds us of Oppenheim and Putnam's system of reductive levels. This idea has come to be known as the *hierarchy thesis* (Khalidi 1993, 1998) and the key challenge to this thesis is the possibility of cross-cutting kinds (e.g., Tobin 2010a).

Some common examples of cross-cutting kinds come from biology and chemistry. In biology, such cases are extremely easy to find. Consider the notoriously tricky kind, *mammal*. Humans are mammals, and so are *monotremes* such as the platypus. The platypus is also oviparous, meaning that it produces offspring by laying eggs, like birds. But birds and humans cannot be classified together either as mammals or as oviparous. So, given the hierarchy thesis, something must give. Proponents of the hierarchy thesis, such as Ellis, may of course give up problematic biological kinds like a mammal. But the problem is that even if we were to rule out biological kinds as real natural kinds, there are examples further down the hierarchy as well. To borrow an example from Tobin, albumin and renin are both members of the kind *protein*, whereas renin and the hairpin ribozyme are members of the kind *enzyme*. However, the hairpin ribozyme and albumin do not fit together either as enzymes or proteins. Even though many enzymes are also proteins, not all of them are protein and hence they are not a subkind of proteins, and nor are proteins a subkind of enzymes (see Tobin 2010a for details; for comparison between biological and chemical kinds, see Havstad 2018). This result has, unsurprisingly, been taken to favour pluralism over monism (cf. Dupré 1995).

There are various ways in which friends of the hierarchy thesis could try to accommodate problematic cases, by modifying or weakening the thesis or by denying that the cases violating the hierarchy thesis amount to real natural kinds. One way to do so would be to argue that when two kinds do overlap in such a way that one is not a subkind of the other, then there must nevertheless be something shared between them, such as a common underlying structure. Tobin considers this strategy, applied to the case of enzymes and proteins:

> It might be argued that the kind *nucleic acid* is the underlying kind of which both enzymes and proteins are composed. Thus, there is a common underlying structure involved. We could certainly subsume the kinds enzyme and protein under the kind nucleic acid and thus claim that [the common underlying structure view] can be supported. However, to do so would be misleading in that the kind nucleic acid masks distinct structural differences between

RNA and DNA. These differences are indicated by the fact of higher-level crosscutting kinds. Thus, to subsume them under a homogeneous grouping would be ontologically misleading. (Tobin 2010a: 184)[23]

Tobin goes on to make the point that reduction is not straightforward in cases such as this – the higher-level kinds here are multiply realised. Indeed, cases such as proteins and enzymes are particularly tricky, because of their complex three-dimensional structure which is responsible for a majority of their biological functions. (See Section 3 for further details on the case of proteins). As Tobin argues, classifying proteins and enzymes strictly in terms of the underlying chemical structure would leave out many important details that we do in fact need in science. This is a point bolstered most recently by Bartol (2016) and Joyce Havstad (2018). However, as I argued in Section 3, it is possible to accept this point – to admit that there are important scientific classifications that cannot be captured in terms of the underlying chemical structure and hence accept semantic pluralism – while also promoting ontological unity. Tobin suggests that in cases such as the one at hand, 'Reduction is precluded by the fact that categories such as albumin and ribozymes are determinable categories' (Tobin 2010a: 184). This is indeed evidence that points towards the indispensability of these categories in scientific practice, but arguably it does not prevent ontological reduction, which is compatible with theoretical or semantic anti-reductionism.[24]

What conclusion should we draw from the case of cross-cutting kinds? For the likes of Ellis, it means that we should abandon many higher-level kinds, even if they are accepted as real classifications. Another approach is to go for *conventionalism* about natural kinds, a view advocated by Hacking (2007b) and dismissed by Tobin. This type of view suggests that classification is a matter of convention or agreement, perhaps based on our pragmatic interests. We saw this type of domain- or interest-relativeness with regard to Slater's account of kinds as well, but the conventionalist could go much further and be explicitly anti-realist about kinds. Instead, the lesson that most realists about natural kinds would draw is one of pluralism, albeit the semantic and ontological varieties of pluralism sometimes seem to be confused here – Khalidi's (2013: 72) take looks like a type of semantic pluralism given that he distances himself from Dupré's apparent ontological pluralism. Tobin's view is more conciliatory, as she argues

---

[23] It may be worth noting that nucleic acid might not strictly speaking be a *component* of all enzymes and proteins, but I take it that Tobin has in mind the key role of nucleic acids in specifying the sequence of amino acids that compose proteins and many enzymes.

[24] Compare with Kincaid (1990), who can be seen as arguing against the semantic reducibility of biology into chemistry, and Gillett (2010: 188), who notes that semantic anti-reductionism is compatible with ontological reduction; see also Tahko (2020).

that an acceptance of cross-cutting categories does not entail that the boundaries between them are arbitrary (Tobin 2010a: 189). So, in her view, one can be a realist about natural kinds while accepting that some cases are cross-cutting. I would agree that this is possible, as the hierarchy thesis itself is by no means definitive of natural kind realism. In fact, even fuzzy boundaries may be acceptable, as Khalidi (2013: 68) argues, and this certainly needs to be the case if one wishes to retain realism about categories such as biological species. Yet concessions at least to *semantic* pluralism may need to be made even in the realm of (bio-)chemical kinds, as the case of proteins and enzymes demonstrates. (For a related point about other chemical kinds, see Hendry 2012.)

So, what is the upshot regarding the unity of science? The results clearly favour *some* kind of pluralism, but what kind? As I hope to have made clear, no one really believes in eliminative semantic unity in the sense that entails the denial of semantic pluralism. But some, like perhaps Dupré (e.g. 1995, 2012), might take the provided results to favour *ontological disunity* and hence ontological pluralism as well. It is not entirely clear just how strong Dupré's sense of pluralism really is, but it is in any case stronger than what Khalidi, for instance, seems to have in mind (Khalidi 2013: 72). In my view, the resulting pluralism here is arguably nothing more than semantic or taxonomic pluralism of the type that we have already discussed in detail. If this is the case, ontological reductionism and reductive ontological unity are not immediately threatened, compatible as they are with semantic pluralism. It is now finally time to outline my preferred account of natural kind monism, which arguably provides a very plausible ontological basis for the unity of science.

4.3 Natural Kind Fundamentalism

I have already expressed my preference for natural kind fundamentalism – the view that natural kinds, as substantial kind universals, constitute one of the fundamental ontological categories (see also Keinänen and Tahko 2019). The main interest for the view in the present context is that it provides an easy route to natural kind monism and thereby ontological unity: if the real or genuine natural kinds are only the kinds that can be associated with a substantial kind universal, then we have a very simple answer to the kindhood question, and hence a straightforward account of the identity-criteria for natural kinds. The properties clustered together in entities that are members of natural kinds do so because those entities *are* members of certain natural kinds. I should note that I understand universals as *instantiated* (sometimes labelled 'Aristotelian' instead of 'Platonic') – that is, there are no uninstantiated kind universals in a Platonic heaven. Rather, the instantiated universals are multi-located where their members are. This does invite further questions, of course. Let me anticipate one of them, regarding the status of kinds that have no actual members (e.g., potential transuranic elements that have not been synthesised). My view is that in such cases, we can often state what it is to be a (member of) the kind, but strictly speaking the kind does not exist, since there are no instantiated universals. But we may nevertheless say that *were* an entity of kind K to come into existence, it would have such and such properties *because* it is a member of kind K (say, a yet-to-be-synthesised transuranic element).[25] This is the case even if no entity of kind K ever comes to exist in the actual world. Now, I would frame all this in essentialist vocabulary – that is, the identity conditions of the members of K state the *general* essence of kind K. But this is something that I will set aside here (see Lowe 2008 and Tahko 2015b for more details regarding the essentialist formulation).

Of course, there are many other challenges that remain for natural kind fundamentalism, such as the question of our epistemic access to the genuine kinds. I cannot hope to settle all these challenges here, and indeed it is not my purpose to defend the view in detail (but see Lowe 2015 and Hommen, forthcoming, for two excellent defences). It might nevertheless be interesting to assess the view against some of the main competitors. One of them is the view outlined by Bird and Hawley (2011), whereby natural kinds are complex universals. On this view, natural properties may be understood in terms of property universals and natural kinds would thus be bundles (i.e., complexes) of such universals. Rather than engaging with this view in detail, let me raise a simple challenge to it: some of the best candidates for fundamental natural kinds do not comfortably fit into this picture. I have in mind fundamental physical kinds such as fermions and bosons, which appear to be definable (or at least distinguishable) in terms of just *one* natural property (namely spin: fermions have half-integer spin whereas bosons have integer spin). If this is the case, then it does not seem correct to understand these fundamental kinds as complex universals. Instead, we could say that there are at least some natural properties that constitute natural kinds on their own. Accordingly, we would lose at least some of the initial coherence of the view defended by Bird and Hawley. More importantly, we would struggle to address the original kindhood question: which bundles of properties are genuine natural kinds?

Bird (2018a: 1420) makes a surprising suggestion regarding the connection between kinds and properties. He proposes that Boyd's HPC theory, which suggests that properties cluster together (non-accidentally) in virtue of causal

---

[25] Note that the 'because' is not a causal notion here; it refers to the source of the identity and existence conditions of entities of kind K.

mechanisms, could be generalised to all natural kinds and not just biological kinds like Boyd himself perhaps intended. On this view, an HPC would be 'a particular species of complex universal' (Bird 2018a: 1423). Now, the usual challenge for generalising the HPC theory to all kinds is that it is compositional in nature: the clustering together of certain higher-level properties is explained by a lower-level mechanism. But when it comes to fundamental physical kinds, there is evidently no lower-level mechanism to appeal to. Bird, however, expands the account by appealing to laws of nature in addition to mechanisms, and he considers this to ensure that when it comes to physical and chemical kinds, the clustering is in fact much more sharply defined. Yet this once again (as I argued in connection to Khalidi's account) seems to reduce (or delegate) the kindhood question to a question about the laws of nature. We may ask: why do laws of nature cluster properties together in the way in which they do? Bird's answer will be tied to his dispositional essentialist account of laws (in addition to the account of kinds as complex universals). My worry with this strategy of explaining property clustering in terms of causal mechanisms or laws, quite generally, is that this turns the explanatory order on its head. If the greatest motivation to postulate natural kinds is that they support inductive generalisations, then it is the fact that natural properties systematically *do* cluster together into kinds that explains why laws of nature hold. But if this clustering is again explained in terms of the laws, then the account is circular. I do not mean to suggest that Bird is committed to this type of circular explanation though – after all, he does have a distinct account of laws, but one that is only available to dispositional essentialists.[26]

My own view is closer to Lowe's (2006) picture, and in particular I adopt the distinction between two different kinds of universal: property universals (which we may understand as natural properties) and substantial kind universals (i.e., the natural kinds), as we saw in Figure 6. Any complete account of ontological categories should say something about how categories are related to laws of nature and Lowe has already developed a sophisticated theory about the relationship between laws and kinds (see also Tahko 2015c). More precisely, Lowe proposes that we should explain laws in terms of the natures of natural kinds (cf. also Ellis 2001). The kind *electron*, for instance, has it as a part of its nature that its instances have unit negative charge. This essentialist fact can then help to explain various law-like regularities, such as the net negative charge of those ions that have extra electrons. Instead of considering laws as relations between property universals (like in the Dretske-Tooley-Armstrong theory, e.g.

---

[26] One possible route for the dispositional essentialist would be to argue that laws and the clustering of properties are both explained by the essences or natures of the properties.

Armstrong 1997: 223ff.), Lowe considers the truthmakers of laws as properties and relations that characterise kind universals. Consider electrons, which have a determinate mass, electric charge, and spin quantum number as their essential properties. What accounts for the clustering of these three property universals as necessary properties of the kind electron? As Lowe (2015: sec. 6) points out, property universals and, say, the Armstrongian second-order necessitation relations between them cannot do the job. The main problem here is that a set of monadic property universals are clustered to constitute essential properties of a natural kind in some of their co-located occurrences but fail to occur together in others. For instance, muons can have the same spin quantum number and electric charge as electrons, but they have a considerably larger mass.

We can omit the details of Lowe's four-category ontology here and just look at the basic structure, which introduces a kind of double primitivism: both natural properties and kinds constitute a fundamental category. Consider Figure 7.

An important feature that we can observe in Figure 7, even though it is of course overly simplified, is that all natural kinds (small circles) are linked to properties (triangles), but not all properties need to be linked to natural kinds. There could also be natural kinds that are linked to just one property, but the property and the natural kind are nevertheless distinct – hence the double primitivism. For instance, some fundamental physical kinds such as perhaps fermions and bosons might be definable in terms of just one natural property: spin. Obviously, such examples are controversial, but it is a benefit of the present picture that the possibility of natural kinds definable in terms of

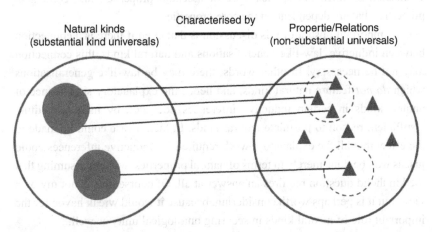

Figure 7 The relationship between natural kinds and their properties (author's own work)

a single property is not ruled out. Although the picture does not illustrate it, there is, of course, nothing that rules out one property being linked to several different natural kinds; the property of spin, for instance, is linked to several kinds. Moreover, the dashed circles on the property side illustrate the ability of the kind to unify the properties – that is, the dashed circle may be regarded as the unification principle in the case of a given kind, be it a causal mechanism or something else. Based on this very simple framework, it is easy to explain why I have been emphasizing that we can make inductive inferences just based on properties rather than kinds. The reason is that there may be properties that are not directly linked to kinds (although they can be entailed by kinds), such as accidents or contingent properties – we may understand the 'loose' triangles to be such properties. A contingent property of a member of a kind may of course be ultimately linked to the kind as well, even if not directly through the kind's unification principle.

To provide an example, consider the diffraction of water waves. Diffraction, the bending of waves around obstacles, is a feature of any wave, but it will of course only be apparent when we have a body of water rather than just one water molecule. Is diffraction a 'core' essential property of the kind water (if it is a kind) or just something that is entailed by the general essence of water? Well, one might think that it is not part of the essence of the kind water that water waves should diffract (even if it were part of the essence of the kind *wave*). If this is the case, we may say that water waves are disposed to diffract, but the core properties of *water* do not necessitate that water waves diffract. Now, all this should be taken with a grain of salt because the example is somewhat esoteric. But we may consider this to be a toy example, the main purpose of which is simply to illustrate the connection between kinds, their core or essential properties, and contingent properties that are dependent on the core properties.

One important upshot of this discussion is that even if there is a connection between inductive, law-like generalisations and natural kinds, this connection may not be necessary. In other words, there may be law-like generalisations which *do not* feature natural kinds, and hence the explanatory significance of natural kinds in certain inductive inferences does not, by itself, constitute a sufficient reason to postulate natural kinds. In fact, a case could be made to the effect that all the explanatory work required for inductive inferences could just as well be done merely in terms of natural properties, without assuming that the kindhood question requires an answer at all. Of course, this is not my own view, but it is perhaps worth considering, because it would wreak havoc for the important role of natural kinds in securing ontological unity as well.

Consider, for instance, inductive generalisations regarding the proportion of black ravens in a sample of ravens or the expectation that two positively charged particles will repel each other, in accordance with Coulomb's law. Could we not explain both cases in terms of natural properties, regardless of whether we consider ravens or even charged particles to constitute a natural kind? In fact, it may be possible to explain at least the raven case without even any reference to *naturalness*, since a random sampling of sufficiently many cases could perhaps be considered a sufficient justification of the inductive generalisation, as Peter Godfrey-Smith (2011) speculates (cf. also Strevens 2012, who uses the raven case as an example). Be that as it may, the case of Coulomb's law is of particular interest due to its wide applicability: it applies to all charged particles. Accordingly, if Coulomb's law features a natural kind, this kind would have to encompass *all* charged particles. On the face of it, the natural kind 'charged particle' seems questionable, even ad hoc. One might think, following Ellis, that such cases have to do with 'global' laws that involve higher-order determinable natural kinds, even if they involve no fundamental natural kinds. Indeed, Ellis himself would have a further possible reply since he accepts 'property kinds' such as 'charge' and 'mass' (Ellis 2001: 23). But these replies are controversial to begin with: Ellis's hierarchical view of natural kinds, property kinds, and the idea of global laws involve heavy commitments and we have already discussed some of the problems of the hierarchical view in the previous section.

So, postulating a natural kind to do the job where simple correlations of natural properties are sufficient – as would seem to be the case with Coloumb's law and charged particles – seems unwarranted, at least if we do it simply for explanatory purposes. Here I would agree with Bird, who states that associating natural kinds with inductive success 'promotes an overly liberal conception of natural kinds' (2018a: 1401). Having said that, explanatory significance may yet have a role to play in this discussion, but, in my view, only to recognize the ontological import of natural kinds. That is, natural kinds may ground the explanatory significance of some law-like generalisations, but not everything that is explanatory has to involve kinds.

There is much more to be said about the exact relationship between kinds, their properties, and the properties that are not directly linked to kinds as well as the relationship between kinds and laws (for discussion, see, e.g., Lowe 2006, 2015, Keinänen and Tahko 2019, and Hommen, forthcoming). It should also be acknowledged that none of this is immediately helpful for addressing the *epistemic* challenge of figuring out which natural properties indeed are linked to natural kinds, but I think some progress can be made if we have an idea about what the ontology could look like – we will leave the epistemic challenge for another day.

In conclusion, what I hope to take from this discussion is simply that natural kind fundamentalism provides one promising way to support my own take on the unity of science, namely a combination of reductive ontological unity and non-eliminative semantic disunity. The toolbox for this combination employs ontological reduction, semantic pluralism, and natural kind monism, the last of which I have just specified in terms of natural kind fundamentalism. This combination is certainly not the only way to defend ontological unity, but anyone who wishes to defend ontological unity does eventually need to give an account of the ontological basis of this unity, which is why a considerable part of this Element deals with the important question of how natural kinds fit into this picture.

References

Aizawa, K. and Gillett, C. (eds.) (2016). *Scientific Composition and Metaphysical Grounding*, New York: Palgrave MacMillan.

Aizawa, K. and Gillett, C. (2019). Defending Pluralism about Compositional Explanations. *Studies in History and Philosophy of Biological & Biomedical Sciences*, 78, 101202.

Antony, L. (2003). Who's Afraid of Disjunctive Properties? *Philosophical Issues*, 13, 1–21.

Armstrong, D. M. (1997). *A World of States of Affairs*, Cambridge: Cambridge University Press.

Bartol, J. (2016). Biochemical Kinds. *British Journal for the Philosophy of Science*, 67, 531–551.

Bird, A. (2018a). The Metaphysics of Natural Kinds. *Synthese*, 195, 1397–1426.

Bird, A. (2018b). I – Fundamental Powers, Evolved Powers, and Mental Powers. *Aristotelian Society Supplementary Volume*, 92, 247–275.

Bird, A. and Tobin, E. (2018). Natural Kinds. In E. N. Zalta, ed., *The Stanford Encyclopedia of Philosophy* (Spring 2018 edition), https://plato.stanford.edu/archives/spr2018/entries/natural-kinds/

Boyd, R. (1991). Realism, Anti-foundationalism, and the Enthusiasm for Natural Kinds. *Philosophical Studies*, 61, 127–148.

Boyd, R. (1999). Homeostasis, Species, and Higher Taxa. In R. Wilson, ed., *Species: New Interdisciplinary Essays*, Cambridge, MA: MIT Press, pp. 141–185.

Breitenbach, A. and Choi, Y. (2017). Pluralism and the Unity of Science. *The Monist*, 100, 391–405.

Carnap, R. (1928). *Der Logische Aufbau der Welt*, Leipzig: Felix Meiner Verlag.

Carnap, R. (1934). *The Unity of Science*, London: Kegan Paul, Trench, Trubner, and Co.

Cartwright, N. (1999). *The Dappled World: A Study of the Boundaries of Science*, Cambridge: Cambridge University Press.

Cat, J. (2017). The Unity of Science. In E. N. Zalta, ed., *The Stanford Encyclopedia of Philosophy* (Fall 2017 edition), https://plato.stanford.edu/archives/fall2017/entries/scientific-unity/.

Chakravartty, A. (2007). *A Metaphysics for Scientific Realism: Knowing the Unobservable*, Cambridge: Cambridge University Press.

Chang, H. (2016). The Rising of Chemical Natural Kinds through Epistemic Iteration. In C. Kendig, ed., *Natural Kinds and Classification in Scientific Practice*, London: Routledge, pp. 33–47.

Clapp, L. (2001). Disjunctive Properties: Multiple Realizations. *Journal of Philosophy*, 98, 111–136.

Daston, L. and Galison, P. (2007). *Objectivity*, New York: Zone Books.

Dorr, C. and Hawthorne, J. (2013). Naturalness. In K. Bennett and D. Zimmerman, eds., *Oxford Studies in Metaphysics*, vol. 8, Oxford: Oxford University Press, pp. 3–77.

Dupré, J. (1983). The Disunity of Science. *Mind*, 92, 321–346.

Dupré, J. (1995). *The Disorder of Things: Metaphysical Foundations of the Disunity of Science*, Cambridge, MA: Harvard University Press.

Dupré, J. (2012). *Processes of Life: Essays in the Philosophy of Biology*, Oxford: Oxford University Press.

Dyson, F. (1979). Time without End: Physics and Biology in an Open Universe. *Reviews of Modern Physics*, 51, 447–460.

Ellis, B. D. (2001). *Scientific Essentialism*, Cambridge: Cambridge University Press.

Fodor, J. (1974). Special Sciences (Or: The Disunity of Science as a Working Hypothesis). *Synthese*, 28, 77–115.

Fodor, J. (1997). Special Sciences: Still Autonomous after All These Years. *Philosophical Perspectives*, 11, 149–163.

Franklin, A. and Knox, E. (2018). Emergence without Limits: The Case of Phonons. *Studies in History and Philosophy of Science Part B*, 64, 68–78.

Gatherer, D. (2010). So What Do We Really Mean When We Say That Systems Biology Is Holistic? *BMC Systems Biology* 4(22), 1–12.

Gillett, C. (2007). Understanding the New Reductionism: The Metaphysics of Science and Compositional Reduction. *Journal of Philosophy*, 104, 193–216.

Gillett, C. (2010). Moving Beyond the Subset Model of Realization: The Problem of Qualitative Distinctness in the Metaphysics of Science. *Synthese*, 177, 165–192.

Gillett, C. (2016). *Reduction and Emergence in Science and Philosophy*, Cambridge: Cambridge University Press.

Godfrey-Smith, P. (2011). Induction, Samples, and Kinds. In J. Campbell, M. O'Rourke, and M. Slater (eds.), *Carving Nature at Its Joints: Topics in Contemporary Philosophy*, vol. 8, Cambridge, MA: MIT Press, pp. 33–52.

Goodwin, W. (2011). Structure, Function, and Protein Taxonomy. *Biology and Philosophy*, 26, 533–545.

Hacking, I. (2007a). The Contingencies of Ambiguity. *Analysis*, 67, 269–277.

Hacking, I. (2007b). Natural Kinds, Rosy Dawn, Scholastic Twilight. *Royal Institute of Philosophy Supplement*, 82, 203–239.

Havstad, J. C. (2018). Messy Chemical Kinds. *British Journal for the Philosophy of Science*, 69, 719–43.

Hawley, K. and Bird, A. (2011). What Are Natural Kinds? *Philosophical Perspectives*, 25, 205–221.

Heil, J. (2003a). *From an Ontological Point of View*, Oxford: Oxford University Press.

Heil, J. (2003b). Levels of Reality. *Ratio*, 16, 205–221.

Hempel, C. G. (1942). The Function of General Laws in History. *Journal of Philosophy*, 39, 35–48.

Hempel, C. G. and Oppenheim., P. (1948). Studies in the Logic of Explanation. *Philosophy of Science*, 15, 135–175

Hempel, C. G. (1965). *Aspects of Scientific Explanation*, New York: Free Press.

Hendry, R. F. (2010). Ontological Reduction and Molecular Structure. *Studies in History and Philosophy of Modern Physics*, 41, 183–191.

Hendry, R. F. (2012). Chemical Substances and the Limits of Pluralism. *Foundations of Chemistry*, 14, 55–68.

Hommen, D. (forthcoming). Kinds as Universals: A Neo-Aristotelian Approach. *Erkenntnis*. doi: 10.1007/s10670-019-00105-6.

Keinänen, M. and Tahko, T. E. (2019). Bundle Theory with Kinds. *Philosophical Quarterly*, 69, 838–857.

Kemeny, J. G., and P. Oppenheim (1956). On Reduction. *Philosophical Studies*, 7, 6–19.

Kendig, C. and Grey, J. (forthcoming). Can the Epistemic Value of Natural Kinds Be Explained Independently of Their Metaphysics? *British Journal for the Philosophy of Science*. doi: 10.1093/bjps/axz004.

Khalidi, M. A. (1993) Carving Nature at the Joints. *Philosophy of Science*, 60, 100–113.

Khalidi, M. A. (1998) Natural Kinds and Crosscutting Categories. *Journal of Philosophy*, 95, 33–50.

Khalidi, M. A. (2013). *Natural Categories and Human Kinds: Classification in the Natural and Social Sciences*, Cambridge: Cambridge University Press.

Khalidi, M. A. (2016). Mind-Dependent Kinds. *Journal of Social Ontology*, 2, 223–246.

Khalidi, M. A. (2018). Natural Kinds as Nodes in Causal Networks. *Synthese*, 195, 1379–1396.

Kim, J. (1992). Multiple Realization and the Metaphysics of Reduction. *Philosophy and Phenomenological Research*, 52, 1–26.

Kincaid, H. (1990). Molecular Biology and the Unity of Science. *Philosophy of Science*, 57, 575–93.

Kistler, M. (2018). Natural Kinds, Causal Profile, and Multiple Constitution. *Metaphysica*, 19, 113–135.

Kitcher, P. (1981). Explanatory Unification. *Philosophy of Science*, 48, 507–531.

Knox, E. (2016). Abstraction and its Limits: Finding Space For Novel Explanation. *Noûs*, 50, 41–60.

Kornblith, H. (1993). *Inductive Inference and Its Natural Ground*, Cambridge, MA: MIT Press.

Kripke, S. A. (1980). *Naming and Necessity*, Cambridge, MA: Harvard University Press.

Ladyman, J., Ross, D., Spurrett, D., and Collier, J. (2007). *Every Thing Must Go: Metaphysics Naturalized*, Oxford: Oxford University Press.

Lange, M. (2004). The Autonomy of Functional Biology: A Reply to Rosenberg. *Biology and Philosophy*, 19, 93–109.

Lange, M. (2009). *Laws and Lawmakers*, Oxford: Oxford University Press.

LaPorte, J. (2004). *Natural Kinds and Conceptual Change*, Cambridge: Cambridge University Press.

Le Poidevin, R. (2005). Missing Elements and Missing Premises: A Combinatorial Argument for the Ontological Reduction of Chemistry. *British Journal of Philosophy of Science*, 56, 117–134.

Lewis, D. (1983). New Work for a Theory of Universals. *Australasian Journal of Philosophy*, 61, 343–377.

Llored, J.-P. (2012). Emergence and Quantum Chemistry. *Foundations of Chemistry*, 14, 245–274.

Lowe, E. J. (1989). What Is a Criterion of Identity? *Philosophical Quarterly*, 39, 1–21.

Lowe, E. J. (2006). *The Four-Category Ontology*, Oxford: Oxford University Press.

Lowe, E. J. (2008). Two Notions of Being: Entity and Essence. *Royal Institute of Philosophy Supplement*, 62, 23–48.

Lowe, E. J. (2015). In Defence of Substantial Universals. In G. Galluzzo and M. J. Loux, eds., *The Problem of Universals in Contemporary Philosophy*, Cambridge: Cambridge University Press, pp. 65–84.

Magnus, P. D. (2012). *Scientific Enquiry and Natural Kinds: From Planets to Mallards*, New York: Palgrave Macmillan.

Mill, J. S. (1843/1882). *A System of Logic* (8th ed.), New York: Harper & Brothers.

Millikan, R. (1999). Historical Kinds and the 'Special Sciences'. *Philosophical Studies*, 95, 45–65.

Mitchell, S. (2002). Integrative Pluralism. *Biology and Philosophy*, 17, 55–70.

Mitchell, S. (2003). *Biological Complexity and Integrative Pluralism*, Cambridge: Cambridge University Press.

Morrison, M. (2000). *Unifying Scientific Theories: Physical Concepts and Mathematical Structures*, Cambridge: Cambridge University Press.

Nagel, E. (1961). *The Structure of Science: Problems in the Logic of Scientific Explanation*, New York: Harcourt, Brace & World.

Nathan, M. (2017). Unificatory Explanation. *British Journal for the Philosophy of Science* 68, 163–186.

Needham, P. (2008). Is Water a Mixture? Bridging the Distinction Between Physical and Chemical Properties. *Studies in History and Philosophy of Science Part A*, 39, 66–77.

Needham, P. (2010). Nagel's Analysis of Reduction: Comments in Defence as Well as Critique. *Studies in History and Philosophy of Modern Physics Part B*, 41, 163–170.

Needham, P. (2011). Microessentialism: What Is the Argument? *Noûs*, 45, 1–21.

Nesse, W. D. (2011). *Introduction to Mineralogy*, 2nd ed., Oxford: Oxford University Press.

Ney, A. (2010). Convergence on the Problem of Mental Causation: Shoemaker's Strategy for (Nonreductive?) Physicalists. *Philosophical Issues*, 20, 438–445.

Okasha S., (2002). Darwinian Metaphysics: Species and the Question of Essentialism. *Synthese* 131, 191–213.

Oppenheim, P. and Putnam, H. (1958). Unity of Science as a Working Hypothesis. *Minnesota Studies in the Philosophy of Science*, 2, 3–36.

Patrick, K. (2018). Unity as an Epistemic Virtue. *Erkenntnis* 83, 983–1002.

Polger, T. W. and Shapiro L. A. (2016). *The Multiple Realization Book*, Oxford: Oxford University Press.

Putnam, H. (1967). Psychological Predicates. In W. H. Capitan and D. D. Merrill, eds., *Art, Mind, and Religion*, Pittsburgh: University of Pittsburgh Press. Reprinted as The Nature of Mental States, in Ned Block, ed., *Readings in Philosophy of Psychology*, vol. 1, Cambridge: Harvard University Press, 1980.

Putnam, H. (1975). *Mind, Language and Reality. Philosophical Papers*, vol. 2, Cambridge: Cambridge University Press.

Rosenberg, A. (1985). *The Structure of Biological Science*, Cambridge: Cambridge University Press.

Rosenberg, A. (1994). *Instrumental Biology, or the Disunity of Science*, Chicago: University of Chicago Press.

Rosenberg, A. (2006). *Darwinian Reductionism, or, How to Stop Worrying and Love Molecular Biology*, Chicago: University of Chicago Press.

Schickore, J. (2008). Doing Science, Writing Science. *Philosophy of Science*, 75, 323–343.

Seifert, V. (2017). An Alternative Approach to Unifying Chemistry with Quantum Mechanics. *Foundations of Chemistry*, 19, 209–222.

Seifert, V. (2019). Reduction and Emergence in Chemistry. *The Internet Encyclopedia of Philosophy*. www.iep.utm.edu/red-chem/.

Sidelle, A. (2009). Conventionalism and the Contingency of Conventions. *Noûs*, 43, 224–241.

Sider, T. (2011). *Writing the Book of the World*, Oxford: Oxford University Press.

Slater, M. H. (2009). Macromolecular Pluralism. *Philosophy of Science*, 76, 851–863.

Slater, M. H. (2015). Natural Kindness. *British Journal for the Philosophy of Science*, 66, 375–411.

Stanford, P. K. and Kitcher, P. (2000). Refining the Causal Theory of Reference for Natural Kind Terms. *Philosophical Studies*, 97, 99–129.

Strevens, M. (2012). The Explanatory Role of Irreducible Properties. *Noûs*, 46, 754–780.

Symons, J., Pombo, O., and Torres, J. M. (eds). (2011). *Otto Neurath and the Unity of Science*, Dordrecht: Springer.

Tahko, T. E. (2012). Boundaries in Reality. *Ratio*, 25, 405–424.

Tahko, T. E. (2015a). *An Introduction to Metametaphysics*, Cambridge: Cambridge University Press.

Tahko, T. E. (2015b). Natural Kind Essentialism Revisited. *Mind*, 124, 795–822.

Tahko, T. E. (2015c). The Modal Status of Laws: In Defence of a Hybrid View. *Philosophical Quarterly*, 65, 509–528.

Tahko, T. E. (2018). Fundamentality. In E. N. Zalta, ed., *The Stanford Encyclopedia of Philosophy* (Fall 2018 edition). https://plato.stanford.edu /archives/fall2018/entries/fundamentality/.

Tahko, T. E. (2020). Where Do You Get Your Protein? Or: Biochemical Realization. *British Journal for the Philosophy of Science*, 71, 799–825.

Tobin, E. (2010a). Crosscutting Natural Kinds and the Hierarchy Thesis. In H. Beebee and N. Sabbarton-Leary, eds., *The Semantics and Metaphysics of Natural Kinds*, London: Routledge, pp. 179–191.

Tobin, E. (2010b). Microstructuralism and Macromolecules: The Case of Moonlighting Proteins. *Foundations of Chemistry*, 12, 41–54.

Tobin, E. (2013). Are Natural Kinds and Natural Properties Distinct? In S. Mumford and M. Tugby, eds., *Metaphysics and Science*, Oxford: Oxford University Press, pp. 164–182.

van Brakel, J. (1986). The Chemistry of Substances and the Philosophy of Mass Terms, *Synthese*, 69, 291–324.

van Brakel, J. (2010). Chemistry and Physics: No Need for Metaphysical Glue. *Foundations of Chemistry*, 12, 123–136.

VanDeWall, H. (2007). Why Water Is Not H_2O, and Other Critiques of Essentialist Ontology from the Philosophy of Chemistry. *Philosophy of Science*, 74, 906–919.

van Riel, R. (2011). Nagelian Reduction beyond the Nagel Model. *Philosophy of Science*, 78, 353–375.

van Riel, R. (2014). *The Concept of Reduction*, Dordrecht: Springer.

Waters, C. K. (2016). No General Structure. In M. Slater and Z. Yudell, eds., *Metaphysics in Philosophy of Science*, Oxford: Oxford University Press, pp. 81–108.

Williams, N. (2011). Putnam's Traditional Neo-Essentialism. *Philosophical Quarterly*, 61, 151–170.

Wilson, J. (2010). Non-reductive Physicalism and Degrees of Freedom. *British Journal for the Philosophy of Science*, 61, 279–311.

Wilson, J. (2011). Non-reductive Realization and the Powers-Based Subset Strategy. *The Monist*, 94, 121–154.

Wilson, J. (2015). Metaphysical Emergence: Weak and Strong. In T. Bigaj and C. Wüthrich, eds., *Metaphysics in Contemporary Physics, Poznan Studies in the Philosophy of the Sciences and the Humanities*, Leiden and Boston: Brill/Rodopi, pp. 345–402.

Wilson, J. (forthcoming). *Metaphysical Emergence*, Oxford: Oxford University Press.

Wimsatt, W. (1987). False Models as Means to Truer Theories. In M. H. Nitecki and A. Hoffman, eds., *Neutral Models in Biology*, Oxford: Oxford University Press, pp. 23–55.

Acknowledgements

I would like to thank the MetaScience team – Francesca Bellazzi, Alex Franklin, Toby Friend, Samuel Kimpton-Nye, and Vanessa Seifert – for many useful discussions on the book's theme, as well as for comments on an earlier version of the manuscript, and Elle Chilton-Knight for proofreading. Thanks also to Carl Gillett for helpful tips on the relevant literature as well as to two reviewers for Cambridge for many detailed comments and suggestions. Finally, thanks to the series editor Jacob Stegenga for helpful comments. The research leading to this Element has received funding from the European Research Council under the European Union's H2020 Programme (ERC-CoG-2017) 'The Metaphysical Unity of Science' project, Grant Agreement no.771509.

Cambridge Elements ⹀

Philosophy of Science

Jacob Stegenga
University of Cambridge

Jacob Stegenga is a Reader in the Department of History and Philosophy of Science at the University of Cambridge. He has published widely on fundamental topics in reasoning and rationality and philosophical problems in medicine and biology. Prior to joining Cambridge he taught in the United States and Canada, and he received his PhD from the University of California San Diego.

About the Series

This series of Elements in Philosophy of Science provides an extensive overview of the themes, topics and debates which constitute the philosophy of science. Distinguished specialists provide an up-to-date summary of the results of current research on their topics, as well as offering their own take on those topics and drawing original conclusions.

Cambridge Elements ☰

Philosophy of Science

Elements in the Series

Scientific Knowledge and the Deep Past: History Matters
Adrian Currie

Philosophy of Probability and Statistical Modelling
Mauricio Suárez

Relativism in the Philosophy of Science
Martin Kusch

Unity of Science
Tuomas E. Tahko

A full series listing is available at: www.cambridge.org/EPSC